The Beastly Book

Also by Jeanne K. Hanson

illustrated by Glenn Wolff

Of Kinkajous, Capybaras, Horned Beetles, Seladangs and the Oddest and Most Wonderful Mammals, Insects, Birds, and Plants of Our World

Also Illustrated by Glenn Wolff

When in France

The Field Guide to Wildlife Habitats of the Eastern United States

The Field Guide to Wildlife Habitats of the Western United States

Outdoors

Even Brook Trout Get the Blues

It's Raining Frogs and Fishes

The Beastly Book

100 OF THE WORLD'S
MOST DANGEROUS CREATURES

JEANNE K. HANSON

Illustrated by
GLENN WOLFF

PRENTICE HALL GENERAL REFERENCE
New York London Toronto Sydney Tokyo Singapore

To the unending, various miracle of evolution.

PRENTICE HALL GENERAL REFERENCE
15 Columbus Circle
New York, New York, 10023

Library of Congress Cataloging-in-Publication Data

Hanson, Jeanne K.
The beastly book : 100 of the world's
most dangerous creatures / Jeanne K. Hanson ; illustrations
by Glenn Wolff.
p. cm.
Includes bibliographical references.
ISBN 0-671-85022-9
1. Dangerous animals—Popular works. I. Wolff, Glenn.
II. Title.
QL100.H345 1993
591.6'5—dc20 CIP 92-40847

Designed by Irving Perkins Associates, Inc.
Manufactured in the United States of America

10 9 8 7 6 5 4 3 2 1

First Edition

Acknowledgments

I would like to thank the following scientists for reviewing the entries for accuracy—J.K.H.:

William Berg (wolves), Minnesota Department of Natural Resources

Brian Joseph (fishes), Minnesota Zoo

Russell Johnson (microbes), University of Minnesota Medical School

Anita Cholewa (plants), University of Minnesota Plant Biology Department

David Withrow (orcas), National Marine Mammal Laboratory, Seattle

Lynn Rogers (bears), U.S. Fish & Wildlife Service

Philip Clausen (insects), University of Minnesota Entomology Department

Dave Noetzel (insects), University of Minnesota Entomology Department

Peter Jordan (moose), University of Minnesota Fisheries & Wildlife Department

Sue Lewis (vampire bat), graduate student, University of Minnesota Ecology & Behavioral Biology Department

Franklin Barnwell (invertebrates exclusive of insects), Ecology & Behavioral Biology, University of Minnesota

Jim Gerholt (reptiles and amphibians), Minnesota Zoo

Mike Don Carlos (porcupine and skunk), Minnesota Zoo

I would like to thank the following people who were extremely helpful and generous with their time in critiquing my illustrations for accuracy—G.W.:

Paul Sieswerda, Aquarium Curator
The Aquarium for Wildlife Conservation

John Behler, Curator, Herpetology,
N.Y.Z.S. The Wildlife Conservation Society

Kristin Hurlin, Artist and Illustrator
Glen Arbor, Michigan

Gary R. Seabrook, M.D., Assistant Professor, Dept. of Vascular Surgery, Milwaukee County Medical Complex

Terrie E. Taylor, D.O., Assistant Professor, Michigan State University and Co-Director, Blantyre Malaria Research Project, Queen Elizabeth Central Hospital, Blantyre, Malawi

A. Christopher Carmichael, Ph.D., Curator of Mammology, Michigan State University Museum

Contents

About These Creatures

This is a book about vicious fishes, hideous insects, major mammals, slinky reptiles, and a few horrifying and very tiny organisms. The accounts of their lives are not only thoroughly accurate but also guaranteed to be safer than meeting these creatures in person. There is a real zoo of havoc here, and I have highlighted the dangers for fun. Together they make a brew that the witches of Macbeth might stir, along with their famous "Scales of dragon, tooth of wolf,/ Witch's mummy, maw and gulf/Of the ravined salt-sea shark,/Root of hemlock digged i'th' dark." They crowd this book in an order based on the whim of zookeeper and brewmeister.

By *dangerous*, I mean distinctly dangerous to human beings. This is a parochial perspective, I realize. But at the risk of sounding incurably "species-ist," it seemed to me the most interesting approach. The dangers included are direct too; I have left out creatures like the locusts who can destroy the grain we need and cause enormous damage—they do it without attacking us bodily.

This is a completely, even a wildly, idiosyncratic collection. There is really no such thing as the one hundred *most* dangerous creatures. After all, that would depend on how many killer bees might sting you, where you might be gored by the rhino, or how full is the venom gland of the snake that might bite. So I sought to present a variety of creatures rather than to attempt an hierarchy of horror. Of course, it would have been possible to fill this book with snakes, sharks, tropical insects. But I think that would have been much less interesting. I have followed my own heart's desire (and fear). The result is that there may be more about cone shells, polar bears and sea anemones, and less about fly agarics and electric eels, than in a book written by someone else. In every case, I've dug to find information that I think will fascinate you and that I bet you don't know, even if the creature is familiar to you.

Not all of these creatures are consistently fatal. Some can be escaped from (the army ant battalions, for example); some are not nearly as dangerous as popular mythology makes them out to be (the black bear, for instance); and some can be merely, though intensely, unpleasant (the horned toad that squirts blood from its eyes when disturbed). Dangerous doesn't always mean deadly. Still, I wouldn't read the whole book at once after dark.

The one hundred creatures featured here are all real. There are no Loch Ness monsters or Big Foots, however wonderful they are to contemplate. Neither are there any beckoning mermaids or "frumias bandersnatches," nor any of the half-horse–half-human centaurs from ancient mythology, nor creatures from science fiction books about far-away planets, though all these creatures would indeed show, even seductively, how closely our human imaginings mingle with the animals. We are all one very real family in this book.

All of these creatures are alive today, too. I have not written about the exquisitely dangerous dinosaurs that I love so well (even though to them I might be an hors d'oeuvre). Living creatures just feel more dangerous, it seems to me. Is that a bear in the underbrush there, a giant leech in the stream?

Of all the dangerous and unpleasant creatures in this book, three are included because people think they are terrible, but they are terribly wrong. Far more dangerous than the wolf, killer whale, and black bear are the fire ant, buttercup, cone shell, and toadfish.

The mystery and fascination of wake-up-and-watch-out-for-it creatures have led me, and I hope will lead you, to a sharper-edged awareness of how we must work much harder to preserve habitats. We need to do more than buy animal T-shirts with big bears and whales on them and read books like this about the marvelous ways of animals. We should vow to use a bit less of the consumer product cornucopia that translates into destruction of these creatures' living places—whether they are forests, or waters, or grasslands.

We can only assume that each of these creatures is an important part of some web within the larger web of nature. It has proved to be true so far. So there must be room on the ark. *Biophilia*, or love for all the forms of life, is somehow built into our bones and we must act on our family feeling. This may seem like a heavy message to place on the head of a deadly nightshade, or a young wolf running in the moonlight, or certainly one of the nasty insects that crawls through these pages. You may not agree with me now that all these dangerous creatures should have places to live. But I hope after reading this book your respect for the creatures of the world will not only be deeper but will galvanize you to some action.

Some of the one hundred dangers here are what scientists call *charismatic megaspecies*, the big bears and lions that have grabbed our attention already to some degree. At the top of their food chains, their continuing existence is proof that these environments are in working order. But others are more modest—the sea wasp and the giant leech, for instance—and their importance is discerned with more difficulty. There may be a few creatures that seem expendable. I doubt that they are.

Dangerous creatures already lie close to the bone in our imaginings. Even more than milder creatures, their power has a beauty. Think for a moment about the totem poles and totem animals of various human cultures. Most obvious are the Northwest Indian tribes' wooden totem poles carved with "killer" whales (orcas), bears, and more, all in bright colors and with big mouths. Remember, too, the old well-wrought European folk and fairy tales that speak of dangerous wolves and more, usually in the woods. The prides of lions, pods of whales, swarms of bees, herds of rhinos, and schools of piranhas all exert a special pull on our imaginations. Wonderfully so.

This book is my totem pole, one so tall with the one hundred creatures I have studied and pondered for so long that it scratches the sky. I respect every creature that I have carved into it and even love them in some way. Within it, I hope you will find your totem pole of fiercely special animals, too. Will yours feature the polar bear, dwarf mongoose, scorpion, crocodile, toadfish, and bombadier beetle in a high pile, with a giant carnivorous land snail plopped on top?

I hope you will delight in hearing about all of these wonderful creatures. Each is another everyday miracle of evolution. Though dangerous, they are wonders of the world.

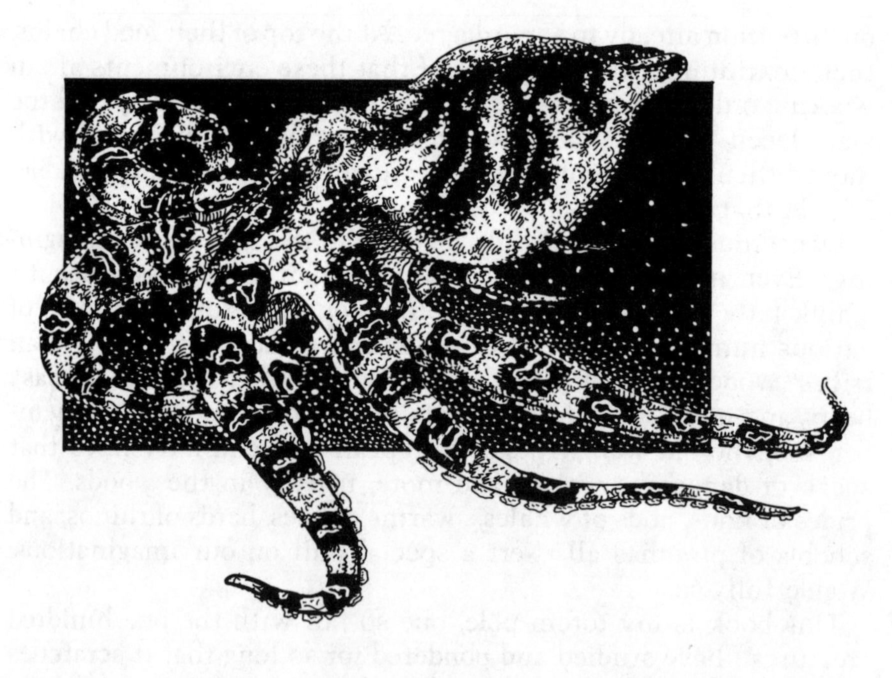

Blue-Ringed Octopus

SMALL, BEAUTIFUL, AND DEADLY

This small and beautiful mollusk is cousin to the nautiluses, cuttlefish, squids, and all the other octopi of the oceans. Swimmers sometimes pick up a blue-ringed octopus to show their friends, setting the bright 8-inch creature on the back of their hand. The blue rings circle its tentacles and make pretty crescents on its purplish brown body. Nobody even feels its little bite. But within five minutes, the bite can cause dizziness and shortness of breath, then its victim's lungs become progressively paralyzed. Within two to three hours, it can cause death. The victim must be put on a respirator quickly and remain there for several hours up to several days. Even then the bite may be fatal.

The way to avoid this seashore fate is to eschew pulling one of these creatures out of an underwater crack or stepping on one in a clump of stuff on the ocean floor. Watch for it off eastern and northern Australia, Indonesia, and the Philippines—areas that harbor one of the two species of blue-ringed octopus. And where there is one creature, there are probably quite a few.

It is not unusual for any octopus or squid to present some danger, since all can bite viciously with the sharp beaks they use to kill their prey and to defend themselves. Their saliva is also venomous to some degree. But the giant octopus, of grade B movie fame, is actually less dangerous than the two blue-ringed species, though a person could get temporarily tangled in its tentacles (up to 32 feet long) if the creature thinks you are food or another octopus invading its territory.

But an octopus figures out a situation fairly quickly. Octopi may be the smartest of the invertebrates and have even been trained to come when called, as well as to distinguish between different-colored balls at an aquarium. Yet more amazing, one study allowed an octopus to watch another octopus choose the "right" ball, whereupon the "student" creature chose the right colors in its tank too. They prefer to eat crabs, mussels, scallops, lobsters, clams, and other such fare rather than to bother people, and they hide in ambush to get their food. They can sometimes figure out how to get their prey out of an unusual container, such as a sealed jar. Octopi are also quite shy. When disturbed by something bigger and less edible, octopi prefer to get away. But remember to stay away from the blue-ringed one.

Vampire Bat

ON A BLOOD DIET

Creatures of the night, these bats do indeed drink blood. Their meal is usually from a cow or a horse or a bird, but one might try to drink human blood too.

Of the more than 900 species of bats, only three are true vampires—the common vampire, the white-winged vampire, and the hairy-legged vampire—and their common diet of blood is unique among mammals (which they are), perhaps even among vertebrates. (Insects such as mosquitoes and ticks engorge with blood too, of course.) Fortunately, these parasitical bats live mostly in the tropics and subtropics of South and Central America, though the common and the white-winged ones live on the edges of the temperate zones in both Americas. They all fly very well throughout their territory and can also run along the ground with great agility.

The common vampire bat causes the most damage, since the wounds it leaves on cattle not only weaken them but allow other organisms to infest them more easily. Its numbers are increasing too. This vampire finds its prey with smell and heat receptors in and below its fleshy nose, and uses echolocation or animal sonar, as well. It attacks the animal by biting most often where its hooves meet the skin but also occasionally on its lips, eyelids, nose, ears, anus, or neck.

These bats occasionally bite people by injecting their razor-sharp incisor teeth into a finger or toe, sometimes a nose or cheek—all accessible and blood-rich body parts.

The small v-shaped wound is kept open with an anticoagulant in the bat's saliva. Through special grooves on its tongue, the bat laps up the blood. If left undisturbed, it will feed for twenty to thirty minutes.

Even after the vampire bat has flown away, the wound continues to bleed because of the anticoagulant. For this reason, a child bitten by two or three bats can die from loss of blood. The bats can also carry rabies and other infectious diseases. The bite itself causes so little pain that the victim is not usually even awakened. (People who must sleep outside in bat-populated areas should try to sleep lightly.) The only good news is that the anticoagulant has benefits. One pharmaceutical company is perfecting it now as a dissolver of blood clots for people in danger of heart attacks.

The common vampire can drink about 15 cubic centimeters of blood per night, which adds up to about one and a half times its own weight. To be able to fly home to its cave or hollow tree, it has evolved kidneys that extract the nutrients from the blood meal and excrete the water very quickly. The bat finds a perch and, within seconds, begins to urinate out the excess water until it is light enough to fly away. Then its kidneys use the remaining water to dilute the pure protein from the blood. Best estimates are that a colony of 100 of these vampires drink enough blood in a year to completely drain twenty-five cows. If a bat goes two nights without blood, it is actually close to starving to death.

To save the lives of their fellow bats, female vampires will regurgitate bits of blood to their dangerously hungry fellow colony members, relatives and nonrelatives alike. To get the full bat to share its blood meal, the starving bat licks its wing and then its lips. Then they hang upside down together, one spitting into the other's mouth. All this goes on in a hollow tree so that we don't have to see it. These bats then live to fly—and try—again the next night. This altruism is relatively unusual among mammals and seems to work because of the bats' communal living habits.

Vampire bats live in groups of eight to twelve adult females and their pups. Each bat has an individual call that the others can recognize. The adult males strive to get the top position in the hollow tree, near this female group, to be at the pinnacle of the dominance hierarchy; the less dominant ones end up sleeping alone, in small groups, or very low in the hollow trees. The females may move among groups every couple of years, but, since they can live up to eighteen years, they end up associating with the same females often. Thus, researchers think a female shares food with another bat because she can expect the other bat to share blood with her the next time. This relationship, which goes beyond the favoring of kin, is called *reciprocal altruism* (quite an achievement for such a much-maligned and scary creature).

Besides sleeping inside a bat-proof building, the best way to keep a vampire bat from attacking you at night is to keep your sleeping area flooded with light. Bats dislike light enough that they stay at home even when the moon is at its brightest. This is not a perfect solution, though, because they sometimes attack their prey on its shaded side. They also like to return to the same wound the next night, so it is possible to put a special anticoagulant on a wound that gradually weakens the bat with repeated feedings, as has been done with cattle.

A final bit of good news: two bat species—the false vampire and the giant spear-nosed—look as though they want blood, but they do not. Watch out for the true vampires, though. And keep in mind Lewis Carroll's little ditty for cheer:

> Twinkle, twinkle, little bat!
> How I wonder what you're at!
> Up above the world you fly,
> Like a tea-tray in the sky.

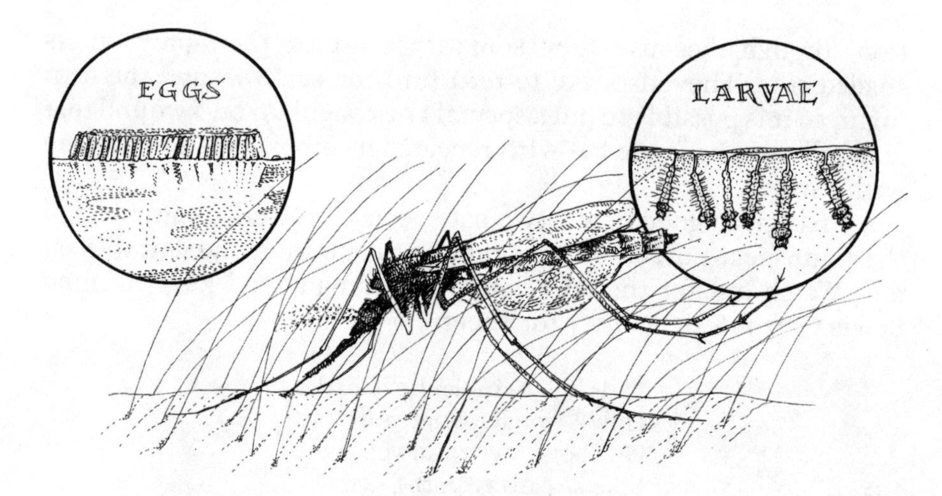

Mosquito

HISTORIC KILLER

"They are a kind of gnat," one observer has written in a state of incipient hysteria, "which consists of almost nothing but legs and sting. They are not seen during the day, but as soon as the sun has set you see them coming out of holes in the ground, or from trees and plants, to torment us, during which time, attacking in large numbers, they give us no peace, by their continual humming of: 'Cousins, cousins, ins, ins,' and biting with their sting so strongly that very often one does not know whether one ought to laugh or cry at the itching which follows subsequently, and which cannot prevent oneself from scratching in such a way that next morning one has nothing but blisters on one's face, hands and all over one's body where they have bitten, so much so that one is unrecognizable." Written in what seems to be a state of grammatical panic (and in a state of partial confusion since mosquitoes are not gnats and do not literally sting), this is part of a journal penned in the late seventeenth century. But anyone who has ever tried to barbecue or quaff a beer outside on the wrong windless American summer evening can certainly sympathize with this sizzling accusation.

Mosquitoes have been buzzing around for 100–200 million years, and they have surely been bothering us since the heyday of the Neanderthals. In fact, the face and body paint used by ancient tribal people may have been partly a defense against their zing. They are probably also the culprit in the death of Alexander the Great, cutting short the expansion of that empire by administering malaria. They stopped Napoleon, too, in his ambition to conquer the New World; he had sent his brother-in-law and a small army to take over Haiti, only to have 22,000 out of 25,000 of the men killed by mosquito-borne yellow fever, and it is said that the conqueror's loss of confidence in French soldiers in hot weather influenced him to sell Louisiana to the United States in 1803. Malaria also killed the intrepid explorer Dr. Livingstone in Africa, we can correctly presume.

The world is still swarming with these powerfully nasty creatures. There are 2,400 to 3,000 mosquito species worldwide, with more than 100 kinds of mosquito in North America alone. The creatures begin their lives as larvae that swim expertly in the still water where the eggs are laid. Nicknamed "wrigglers," they are at this stage nearly transparent, with bristles on their sides and rear ends; with strong mouthparts, they chew up bits of bacteria, yeast, protozoans, and whatever small things are floating or swimming nearby. Next comes the pupal stage, during which time mosquitoes are called "tumblers," since they tumble through the water in comma-shaped somersaults. Soon enough, they become the adult mosquitoes we know and do not love.

With their wraparound compound eyes, they can see in almost all directions at once. Flying toward us on little wings that beat so fast they are more a fibrillation than a muscle action, mosquitoes can hear with tiny ears, called Johnston's organs, found at the base of their antennae. The males use these "ears" mostly to track the female by her wing sounds, whirs of about 600 beats per second. In fact, males will cluster around a tuning fork of this frequency, clog into a motor with this vibration, and even fly into the open mouth of a human singer foolish enough to hold this note long enough.

Their sense of smell is just as good. They are attracted to carbon dioxide, lactic acid, ammonia, water vapor, and warmth, using all these sure-fire signs of a warm-blooded creature to zero in on large prey, including us. A change in heat of a small fraction of a degree, indicating the presence of a warm body, is enough for them to notice. What they are looking for, of course, is blood. Female mosquitoes need a blood meal to be able to bear their eggs (though both sexes also eat flower nectar too). The males don't need blood and therefore don't bite.

The bite is an orchestration of legs, two little lances, two tiny saws, a blood tube, and an anticoagulant. Also, later, an itch.

It is not just in the faraway tropics that mosquitoes are dangerous. Worldwide, more than 2 billion people live in malaria areas, contracting more than 200 million new cases every year. Three new antimalarial drugs are now being tested by the World Health Organization, since the malarial parasites have become resistant to drugs that were previously effective. There are mosquito species carrying malaria now flying over some of California's rice fields and a couple of cases of the disease have even been found in New Jersey. Another species that carries dengue fever throughout Africa has moved into eighteen states in the southern United States. And a newly arrived species of mosquito is beginning to transmit a rare but usually fatal kind of encephalitis in the southeastern and midwestern United States, especially in Florida.

Throughout history, mosquitoes have actually killed more people than all of the wars put together. They deal their death through a medley of diseases such as malaria, yellow fever, dengue fever (a painful but not usually fatal fever that affects the bones), filariasis (which leads to elephantiasis), encephalitis (a swelling of the brain that can be fatal), and even dog heartworm. The death rate is estimated to be at least one million people every year. And, each year, they initiate more than 250 million new cases of illness, which probably makes them the world's most dangerous creature of all.

To do mosquitoes in is difficult. Special lights and sonic repellents do not work, researchers say. Killing the larvae in the water with insecticides or bacteria is probably the most effective way, as is cutting down on their breeding places by eliminating standing water (even rainwater in old tires). Eating a lot of garlic actually seems to help, on a personal level. So do the newest mosquito nets (becoming available in Africa) that are treated with insecticides against the malaria-carrying species. Yet more research has been successful in deriving a new combined vaccine that, so far, has protected experimental mice against fatal malaria. There are also studies under way that use the mosquito's natural predators against it, as well as studies analyzing mosquito molecular genetics to make them less able to carry diseases. Our own slaps and nostrums have hardly been effective.

Candiru

STRANGE PARASITE

Even readers who feel an affinity for crocodiles and piranhas will want to avoid swimming in the Orinoco River of South America and in the Amazon in general. There lives the tiniest "sea monster" of them all. Only about an inch long and almost transparent, the candiru is a parasite. (This is unusual among fish.)

The candiru sucks blood. And it generally finds it by swimming right into the gill chambers of larger fish. Though it has small, dark eyes, it operates mostly by smell. When it sniffs fishy blood in the water, it swims against the current and into the other fish's body. After feeding, it swims out again. In between meals it hides in the sandy river bottom.

From our nonfish perspective, the problem is that it sometimes gets confused and swims into the body orifice of a person. A good way to avoid having a candiru for an intimate visitor is to not urinate in the water. (This is just what your mommy already told you.) The current created is apparently quite similar to the one around a fish's gill covers. And, of course, a temporary entrance into the body has been opened.

Once inside you, the candiru's little spines point toward the interior, making it impossible to pull out. Since it cannot swim out itself (your body is not shaped like a fish's gill), it remains inside. When it reaches the bladder, it will give you fatal blood poisoning. So victims of this little fish must literally choose between death and amputation of the organ this creature entered.

Komodo Dragon

WORLD'S LARGEST LIZARD

Named for the desolate island of Komodo in Indonesia, these giant, fearsome-looking lizards grow up to at least 10 feet long and can weigh in at 500 pounds or more. The largest lizards in the world, they can easily kill a water buffalo, wild pig, deer, goat, another Komodo dragon, or person (which they have been known to do). Smelling prey almost 4 miles away if the wind is right, they can outrun the fastest person, as well as swim, climb, and dig.

Komodo dragons are ugly. Dark gray and scaly even on their legs, they have flabby necks, chunky snouts, very sharp teeth, and very dull-looking eyes. They are nearly deaf and cannot see very well. They are also cannibalistic when the going gets rough, even of their own family members, thus forcing the young to live up in trees for their first two years.

Komodo is a volcanic, lunar-looking island about 400 kilometers east of Bali that also features flying foxes and a ring of coral reefs. It can be reached only by a long ferry ride twice a week from several neighboring islands. Timed to the ferry arrivals, park officials on the island create a goat-eating frenzy among the dragons. Tourists are herded into a safe enclosure, then the rangers lead a goat out, slit its throat, and cut off one of its legs. The dragons are now going wild outside the fence. The guide throws the dead goat to them, and tourists get to see and photograph them clawing over each other to circle the carcass, then tear at it for about fifteen minutes. (The dragons, usually solitary, get together in the wild to share a kill too.) Ready for the finale? The rangers throw in the remaining goat leg to see which dragon can get it first. Blood and goat meat hang from the dragons' mouths.

There are signs posted on the island's paths warning people not to hike anywhere without a guide. Stories are told of one Swiss visitor who ignored the warnings: only his hat was found. They have also dug up and eaten human corpses.

Komodo dragons live only on this island and five other small Indonesian islands. Islands are like arks and can preserve unusual creatures, so the dragons have been lucky here. They are rare and endangered and have finally been bred in captivity outside Indonesia, only in this decade. Look for Komodo dragons at the Washington Zoo.

Poison Dart Frog

THE EPONYM

Small and brightly colored, these frogs forage in the jungle leaf litter all day. Many live high in the forest, inside water-filled aerial plants called bromeliads. After elaborate courtship, just four to six eggs hatch in some of these poison dart frog species. The tadpoles wriggle right onto the back of one of their parents for protection. And, in some species, parent frogs even feed their new babies some of the special infertile eggs.

All species in this family of small frogs have glands on their skin with particularly nasty toxins. A single frog may pack enough poison to kill 100 people. And three of the species are used by Amazonian Indians to treat their blowpipe darts (sharp darts blown out of a pipe for hunting) or even, in lesser amounts, to rub their wounds. In the case of two of these three most dangerous species, the Indians skewer the little frogs to sticks and warm them over a fire until they secrete enough of their toxin, which is then used to arm the dart. In the case of the third species, the most toxic of all, the Indians simply pin the frog to the forest floor with a stick and rub their darts over the creature's back. These are frogs that no one would want in a terrarium.

Brown Rat

VERY SOCIABLE

The brown rat is not the largest rat in the world. Of the more than 400 ratty species, that prize goes to the cane rat; at 20 pounds, it eats anything it wants to, and then is sometimes eaten itself, prized as a gourmet delicacy in parts of Africa. But the brown rat, a mere 9 inches long (including tail) and weighing up to 1½ pounds, is the most common rat and has caused the most damage.

Brown rats are also called Norway rats, although they originated in Asia. These creatures colonized Europe and Africa, then settled the New World right along with us. First came their colonization of the East Coast cities, well under way by the late eighteenth century. Then these rats went west right behind the pioneers' covered wagons, some reaching California by the 1850s, and others staying everywhere in between. They have, in fact, followed people everywhere, migrating all over the world, and they now thrive everywhere humans do. They have pretty much taken over the world from the black roof rats, who now live mostly in port cities.

There are at least as many brown rats in the world as there are people. And they have introduced to human beings at least 30 diseases, including rabies, salmonellosis, a couple of kinds of typhus, trichinosis, and bubonic plague. In the latter case, they are helped, of course, by the flea, which transmits the disease from rats to humans. Since rats eat grain, they have also played a part in starving people to death. In the tropics, for example, they eat about a third of the food produced. Unfortunately, not only do they eat human food but they mess it up. Each brown rat creates 50 droppings and releases 3 to 6 quarts of urine every day of its life in grain elevators, the backs of grocery stores, and the corners of restaurants. They shed hairs into our food too. In the United States, they cause about 4 billion dollars worth of damage, most of it to our food in one way or another.

The rat-to-person ratio of one-to-one also applies in the United States. Rats are commensal, that is, they like to eat pretty much what you do—they love chicken and chocolate, for example—as well as any garbage. They will even resort to alcohol when placed under stress (at least in laboratory settings, where they can get it easily). We don't see all of these rats because they live in places where we don't go (garbage dumps, underground burrows, and sewers, for example) and come out mostly at night.

Rats are survivors. Some have been known to survive a flush *down* the toilet, while many come up through it. Others have been seen swimming up to half a mile and treading water for at least two days. Since they are light, they can fall five stories and still live, and since their bodies are rather elastic, they can squeeze through small holes. They can chew through wiring, the walls of houses, and grain elevators, even pipes. (This wiring fetish can lead to fires.) They can bite sharply when disturbed, and rat bites are on the increase in New York City right now. On the Pacific Islands where nuclear weapons were tested and radioactivity levels were high, did the rats die? No, their life span actually seemed to increase. And they have become resistant to some of the strongest poisons.

These almost-indestructible creatures can live in colonies of a few to several hundred rats, where they learn to recognize each other as individuals and fight only with intruders occasionally. They groom their fellows sociably. They communicate well with sounds—usually too high for us to hear—to ward off attacks, express pain, warn each other about poisoned food, and signal the end of a sexual encounter. If one pair and their offspring were left totally undisturbed, they would create 20 million rats in just three years.

Although rats have a reputation for being aggressive, this seems to be true only when they cannot flee. And, contrary to myth, aggression does not only occur between adult male rats and young rats.

These creatures are not dumb at all. When they encounter a new food, they will not eat it for several days, then a few of them may eat a very small bit of it. If a rat gets sick, the others will avoid that food completely. For this reason, rat poisons have to be either slow-acting or must look and smell like anything *but* food. There are a couple of such poisons, one of which gives rats heart attacks. Rats remember what a trap looks like for months, too. And they are so well socialized that if good food becomes scarce, the larger males do not grab all of it—both sexes have equal access.

Remember one thing: bred from the brown rat are those cute little white laboratory rats. And they may well have helped to save more lives than the brown rats have killed.

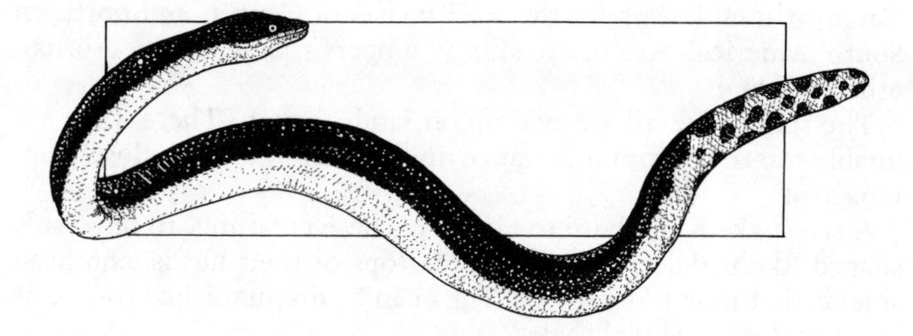

Sea Snakes

SWIMMERS' ALERT

Once upon a time, there were many tales about sea monsters that roamed the oceans. Homer spoke this way about the beautiful Scylla, turned to a monster of the sea: "She hath 12 feet all dangling down, and six necks exceedingly long, and upon each a hideous head, and these three rows of teeth set thick and close full of black death." Yet not this Scylla, nor even Neptune, lord of the sea and all its creatures, could control what may be the most dangerous family of real sea monsters: the fifty-plus species of sea snakes.

Most of these species are less than 4 feet long, though some are 9 feet. Like all snakes, they are usually mild-mannered—they have few enemies and so use their venom just for catching food. But they will bite if disturbed, and some sea snakes pack poison stronger than a cobra's. They are cousins to the cobras in fact, and almost all sea snake species are venomous.

The trick is to avoid stepping on one in shallow, murky water. They prefer the warm shallows at the mouths of rivers and swamps, places where there is no strong surf or strenuous current. They also congregate in large numbers on coral reefs to mate, sometimes whole rafts of these groups rise to the surface, and two species even live in lakes. The sea snake's main home is the tropical water around Southeast Asia, Australia, and sometimes New Zealand and even the open Pacific. There are none in the Atlantic yet, but the yellow-bellied sea snake may soon cross over through the Panama

Canal; it now lives near the west coasts of Central and northern South America. Attention skinny-dippers: some are nocturnal, others diurnal.

The snakes are all washed up on land, though. There, they are unable to strike from a distance and bite hard only if closely approached.

A sea snake is not hard to identify. These creatures all have tails shaped like paddles, nostrils on the tops of their heads, and head shields that look like crowns. They can be distinguished from eels because they have real scales.

Sea snakes are well adapted to their watery lives. Able to stay under without breathing for between a half hour and two hours while active, they can absorb enough oxygen through their skins to stay down for much longer when at rest or asleep. They bite at fish, eels, and prawns when the prey can be trapped in or against a reef or other structure; this allows their poison time to take effect, whereupon the snake pulls the meal into its mouth, usually head-first.

They will also take a baited hook—and some kinds are even prized as food in China, Japan, and sections of Polynesia. They are collected for their skins in parts of Asia too, then turned into belts, shoes, and purses. This has, in fact, put too much pressure on their population in certain areas.

When caught in nets, sea snakes are usually not particularly dangerous, but a few fisherpeople in Asia have died from this slinky surprise. The sea snake's characteristic bite leaves tiny puncture points, and sometimes a few of its teeth are left behind in the wound. Survivors could collect these for a little necklace.

Polar Bear

THE LARGEST OF ALL LAND CARNIVORES

It may look like a stuffed animal, furry to the bottoms of its feet, but it can charge its prey at 25 miles per hour, then swoop up 500 pounds of meat with one paw. It can hiss and roar but is usually very quiet, the better to lie nearby in ambush or to sneak up softly and pounce hard. The Inuits call the polar bear "he who is without shadow."

At 10 feet and between 1,000 and 1,600 pounds, the polar bear is the largest of all land carnivores. Evolved from a small group of coastal brown bears in the high north, probably only during the most recent Ice Age (of the last 2 million years or so), it may well be the newest species of mammal on Earth.

These brawny bears have broken into people's houses right through their front doors (terrifying residents) and also sauntered into the backs of restaurants. So readers in the Arctic are advised to live far from the dump and to avoid keeping a lot of luscious home garbage or even dog food around. And watch the dogs themselves, since polar bears will eat them when other pickings are poor. Of course, the best bet of all may be to stay south of James Bay (the southern part of Hudson Bay), away from the edge of their natural range, where the worst polar bear problems have occurred. In Churchill, Manitoba, the town that has gained fame from television shows about polar bears, the bears appear regularly at the town dump.

This bear's true bread and butter is neither garbage, nor people, nor pets. Ringed seals and bearded seals are its favorites, especially their blubber, skin, and internal organs. To catch these, the bear is willing to drift hundreds of miles from land on ice floes as it looks for basking seals, to swim hard over open ocean using its partly webbed front paws, to float like a big white bobbin, and to lie on the ice for hours above seal holes in the dead of winter. The seals come up to these holes to breathe or to haul out and rest, only to find a big white surprise. A polar bear can smell a piece of blubber 20 miles away. The eyes of the polar bear are sharp for night hunting, and it can even dive after and catch a sea bird that tries to get away under

water. The bears plan their year around the migrations of the krill, which are eaten by the fish, which are eaten by the seals—which are then eaten by the polar bears.

At the end of winter, ever the tricky hunter even when large snow drifts pile up around the seal holes concealing the breathing seals, the bear will claw the snow away and then replace some of it so that the seal will not suspect ambush. Pounce! Then as spring comes and seals build nesting areas on top of the ice, the polar bears can rise up on hind legs and crash right through the den roof to get the baby seal inside (the poor little thing was probably resting while its mother was out fishing). When sneaking up on an adult seal on the bare ice, as they sometimes do too, the polar bear knows enough to cover its only uncamouflaged part—it will hold one white paw over its black nose.

Only "off season," when the ice is motley so the seals are harder to find, does the bear cruise the land for its alternative diet of garbage—as well as its more normal fare of porcupines, muskrats, Canada geese and other birds, with big blueberry bushes for dessert. The animals live mostly off their fat, resting after months of seal hunts.

Polar bears fear only killer whales and walruses, though Inuits in the Northwest Territories have seen a bear occasionally brave one of the latter. It picks up a chunk of ice and bashes in the head of a basking walrus, which it eats for dinner. The black and white whales, however, sometimes choose the bears themselves for a giant white, though furry, feast.

Polar bears are circumpolar creatures, citizens of all the nations that circle the North Pole, and they have been seen as far as 88 degrees north latitude, just 2 degrees from the pole itself, though they usually remain farther south. The males are usually active all year long, but the females, once mated at age 5, den up over the winters to have their cubs, usually twins. In their more southerly haunts, these females will also occasionally den in both fall and spring, possibly to keep cool, and can be seen burrowing not only into the snow and ice available but even into dirt and permafrost. Baby polar bears stay with their mothers until about age $2\frac{1}{2}$ in the northern parts of their range, and about age $1\frac{1}{2}$ farther south (where the seal hunting appears to be easier for a small bear).

Like other species of bears, polar bears have an unusual way of adapting to their dormant periods, perhaps also used in times of food scarcity. They recycle their own urine inside their bodies.

Although animals who are true hibernators must shut down their entire physiological processes to avoid poisoning themselves with their own wastes, these bears have evolved kidneys that can do the recycling. They thus glean the advantages of being incomplete hibernators—they can bear their cubs in the wintertime yet rest enough to conserve fat for nursing. Some scientists think this natural recycling also yields protein, enough to actually sustain the bears through bleak periods when they can't find food and would otherwise go hungry.

Readers who are highly motivated to see polar bears up close can venture to Churchill, Manitoba, a long trip from Winnipeg. The polar bear season here is mid-October to early November. Bears have been coming to this high northern shore for at least as long as written records go back; in the winter of 1619–1620, a Danish explorer seeking the Northwest Passage saw them here. He, alas, had no access to the tourist snow buggies used today. One can take them out onto the ice for one day or several days at a time, while the expedition leaders put out food and the bears plod over. These vehicles stand very high off the ground on huge tundra tires so that the bears cannot break the windows and join the intrepid (but not that intrepid) travelers.

No one is allowed to get out of the tourist buggy. But one scientist I know has gone out onto the ice often at Churchill to study the polar bears much more directly. His only weapon, besides perhaps an immense measure of foolishness, is a squirt gun filled with cayenne pepper and water. Once or twice he has had to squirt it in a bear's eyes, causing temporary itching—and his chance to get away. This scientist has observed that, while these bears may seem to be lolling around like blobs on the ice, they can actually move as fast as lightning.

His advice: "Never turn your back on a polar bear."

Leech

THE CREEPIEST

Who can forget Humphrey Bogart in *The African Queen*, wet from the river with large leeches grabbed to his stomach and arms. . . . Whether applying salt to them really makes them release their grip, as in the celluloid ministrations of Katharine Hepburn, is still not clear; but it is safe to say that most of us share Bogart's aversion to these slimy creatures. However, a few intrepid scientists swim against this stream, purposely seeking out leeches. Standing in a wet marsh in French Guiana several years ago, one scientist found the prize he was seeking for his research: the elusive giant leech (*Haementeria ghilianii*), largest in the world at about 18 inches of green-brown flesh. When young, it feasts on rabbits and turtles. But, contrary to local stories of its ability to suck out the blood of creatures as large as people, the scientist found that this leech's bite is definitely not fatal (just emphatically, pronouncedly, intensely unpleasant). Also, it is difficult to get off.

Picture the leech, large or small, at the water's edge, this segmented worm evolved from the earthworm. Anything making waves? If so, it will quickly swim in that direction by undulating gently through the water. The sucker-like mouth finds, then explores, the object, and the leech may also need to crawl upon it, to find the best warm place to bite. Leeches bite at warmth, since that is often an excellent way of finding their favorite food, which is mammal blood. (The giant leech-seeking scientists mentioned above discovered this by heating the prey and the leeches' lips in experiments.) Their "three serrated jaws create distinctive marks reminiscent of the Mercedes-Benz emblem," say the same incorrigible researchers. A leech meal can be as much as nine times its empty weight in blood, a meal which can last for hours.

The 650-some species of leeches in the world provide plenty of other surprises and even a few redeeming features. Setting aside the giant leech for a moment, let us focus on the leech species used for medicinal purposes, namely blood letting and blood drinking. Historically, this leech has been extraordinarily popular. It was used so

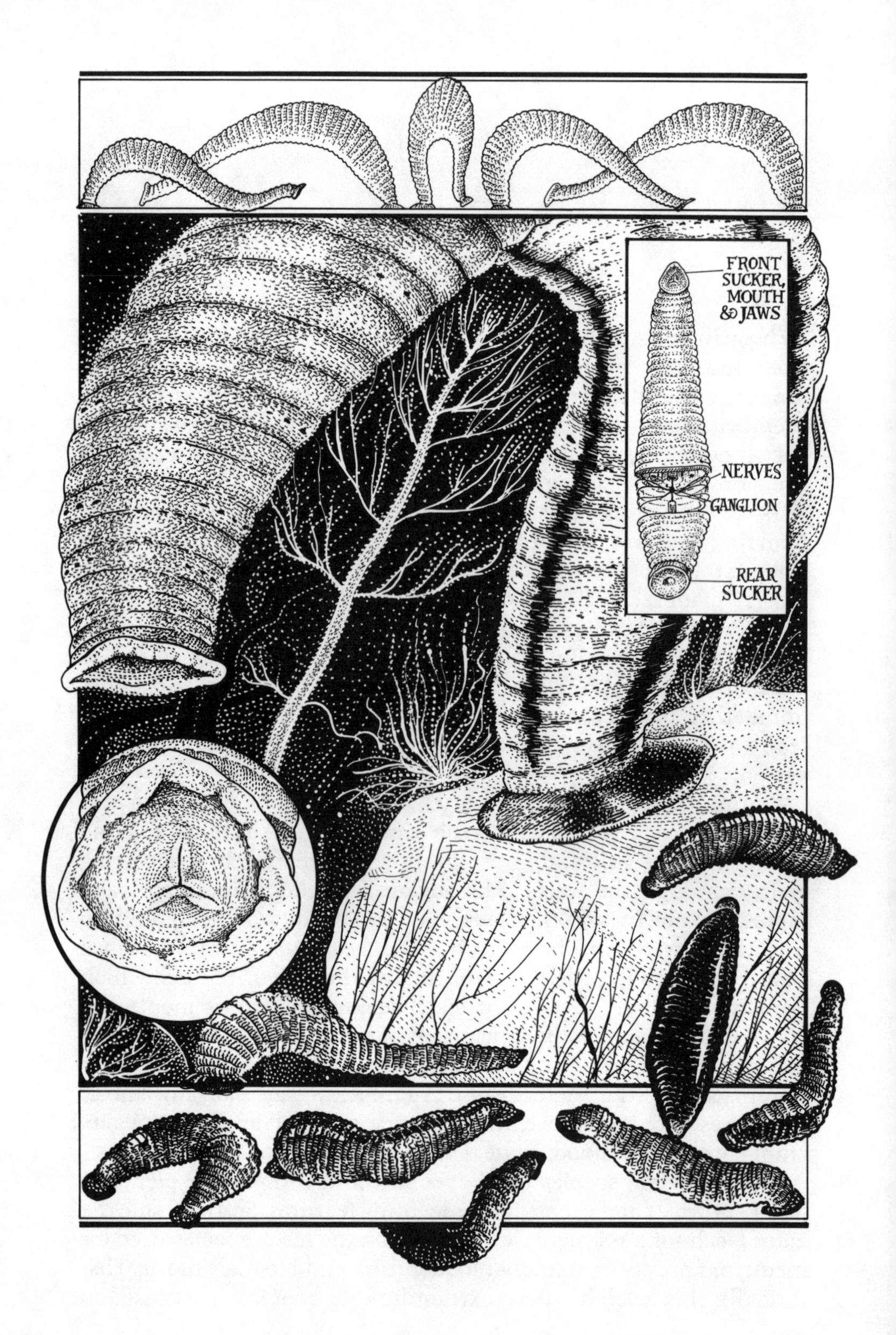
FRONT SUCKER, MOUTH & JAWS
NERVES
GANGLION
REAR SUCKER

often by doctors from the days of second-century Greece through the early nineteenth century that its numbers in Europe began to drop precipitously. For a while, physicians in France, for example, were ordering up to 30 to 40 million a year.

This medicinal leech is popular again today in both Europe and the United States (though not so popular as to make it an endangered species). It is actually being bred for sale by one pharmaceutical company. Its contemporary job is to consume the excess blood under the skin after many operations, from plastic surgery to the reattachment of body parts. This allows body tissue to regenerate better, with less swelling and bloodclotting.

At least some nonscientists love these medicinal leeches too. Some people have been known to keep them in "leech barometers" to foretell storms. When such a weather system is approaching, it is said that the leeches climb up in their glass tubes and move around with great alacrity. But the prize for leech-love probably goes to one hobbyist, a lord in Old England, as reported by a friend of his: "He told us [that] he had [gotten] two favorite leeches. He had been blooded by them last autumn, when he had been taken dangerously ill at Portsmouth; they had saved his life, and he had brought them with him to town, had ever since kept them in a glass, and himself given them fresh water, and had formed a friendship with them. He said he was sure they both knew him and were grateful to him. He had given them different names, 'Home' and 'Cline' [the names of two celebrated surgeons], their dispositions being quite different." One wonders if this friendship involved any regular exchange of bodily fluids. And one marvels again at the eccentricity that seems to grow only in English soil.

The bloodsucking ability of leeches is only one of its potential medicinal uses. Neurologists are using them as an important experimental animal because their nervous systems—so simple and with large moving parts—illuminate those of other of the planet's creatures, including humans. All species of leeches have just 32 ganglia (clusters of nerve cells), each of which directs part of the body via just two paired pathways. And each nerve cell cluster holds just about 400 neurons (the brain cells only in this case). The electrochemical transmitters that allow the leech neurons to communicate are similar to those in the brains of mammals; and one of the transmitters (serotonin) is particularly similar to that in the brains of mollusks, insects, other creatures, and humans. This particular transmitter directs the leech's feeding behavior. In addition, leech saliva includes a strong anticoagulant that keeps its victim's blood flowing. Called *hirudin*, its gene has been cloned and is to be tested in tumor suppression.

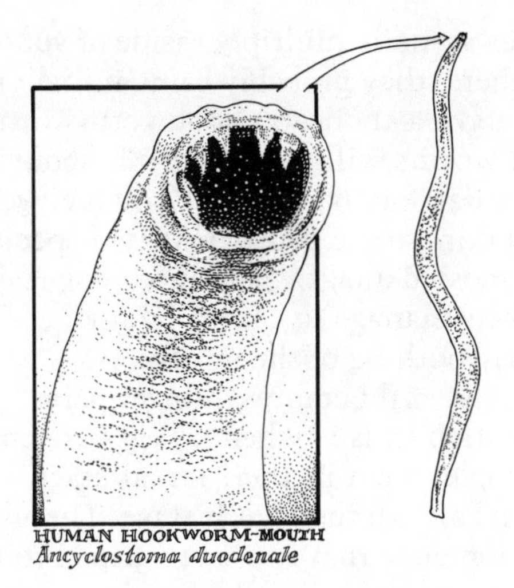

HUMAN HOOKWORM-MOUTH
Ancyclostoma duodenale

Helminth

INTESTINAL WORMS

You would probably rather take your chances with a worm than a tiger? Well consider choosing the tiger. Today, intestinal worms are living inside more than one-third of the world's people. This helminth family, in fact, causes almost as many diseases as there are people on the planet—because, though some people have none, others have more than one disease from the worms. Three members of this worm family alone cause about one billion infections apiece. In one decade, about 54 million helminth infections occur, including some in the United States.

These worms come in three groups: the nematodes (roundworms), the slightly-less-obnoxious trematodes (flukes), and the cestodes (tapeworms). All three groups of worms are common and can grow big. They have complex life cycles and, for part of their lives, this includes moving around inside of humans. All are visible to the naked eye, even in the smallest days of their lives. They get inside of us when we eat their eggs in meat or vegetables, or when we are bitten by insects who are carrying them, or when their larvae actually penetrate our skin. Once inside and mature, they can deposit their eggs in their favorite internal organ. Very few of the helminth species actually multiply inside of you—so, if you have a lot of them in there, they probably came in individually.

Suppose you have a thousand hookworms inside you, which is not rare. These worms will need to drink about 100 milliliters of your blood every day, enough to make you feel weak. Getting rid of them requires taking toxic drugs. Just two species of these hookworms do the most damage, usually in tropical and subtropical areas. Hookworms manage to cause about 60,000 deaths each year worldwide, to say nothing of the infections.

Tapeworms cause mild cramps, hunger pains—guess why—and pain upon defecation (this is when its eggs are coming out of you). They live in people when in their larval stage—making cysts on their organs—and also in their adult stage. The adult tapeworm has a head and suction cups that it uses to grip onto the intestines. Its only other salient organ is a double-sexed region, which lays eggs that must come out (see above). The adult helminth tapeworm can grow up to 30 feet long . . . and, yes, there is plenty of room in the human intestines.

To avoid getting tapeworms, don't eat meat or fish with their cysts on it (it will look "measled"); if you do, you will have an adult worm inside you in about two months. Meats and fish like these have been noted in parts of eastern Europe, Central and South America, Spain, Portugal, parts of Africa, China, and India. The pork tapeworm alone causes 50,000 deaths around the world each year.

Scorpion

ANCIENT CREATURE, NASTY STING

If you find yourself in the deserts or tropical forests of America, Africa, or Arabia, don't pick up a rock too quickly, and shake out those shoes and clothes carefully in the morning. Why? These places are burrows for scorpions during the daytime, at the end of an active night of stinging prey into submission for dinner. And they will not be too tired to sting you the morning after.

All the 1,500-plus species of scorpions are venomous, though some cause "only" intense pain and just 25 species can deal death to humans. Arachnids, they are related to spiders, which they also eat. Some scorpions eat other scorpions too, including, occasionally, their own family members. The largest species kill frogs and small rodents, then strip off their flesh and grind them up. Unlike bees, for whom stinger and venom are defensive, scorpions use theirs quite offensively. The prey is found when the scorpion hears it, at night as well as by daylight.

They carry out these antics—and more—not only in deserts, where popular imagination always places them, but in forests of every warm temperature, savannahs and grasslands, even high on cold mountains and deep into chilly caves (the latter species, blind and pale, are a feature of some of the longer tours in Mammoth Cave, Kentucky). More species are being discovered regularly!

This creature has not changed a great deal since the Silurian era of 350 to 400 million years ago. It is related to the even more ancient sea scorpions of those halcyon days that swaggered through the seas and were up to 2 meters long. Probably some of the years since then have been spent perfecting their venom recipe. That of the buthid scorpion, for example, contains a particularly good array of different amino acids and is being used to study the transmission of chemical messages. Scorpion venom is whipped out of the tails via the stingers and is neurotoxic, meaning it acts on the central nervous system of the prey. Its dinner becomes quickly paralyzed. Digestion is not pretty either—the scorpion spits enzymes onto its prey to digest it externally, then slurps it up.

The family life of scorpions is unusual. Mating is treacherous,

and even more so for the male, who is smaller than the female. He first grabs her pincers shut, then they dance around, tails entwined and with him sometimes stinging her. After a while, he finds a nice stick onto which to deposit his sperm and drags her body over it so that she can absorb it through the proper orifice. She gets the last laugh, though, because while the male tries hard at the end to get away, she tries to eat him up (extra food energy for bearing the babies, you know). Her plot is successful at least 10 percent of the time.

Scorpions usually don't mate until they are about seven years old, and pregnancy is long too—about a year and a half. This may seem extraordinary, but it is indeed matched by their unusually long scorpion lives of fifteen to twenty-five years. Longevity seems to come courtesy of their very slow metabolism, useful for surviving in harsh conditions where water can be scarce and food is available sometimes only once a year; such slow living is not usually characteristic of creatures this small, who generally rush through their life spans on fast metabolisms.

The baby scorpions take their time to grow up. After staying inside—benefiting from a maternal food supply line that bears an amazing resemblance to the placenta of a mammal—they climb onto their mother's back for another two to six weeks of maturation. That back ride is on a tough exoskeleton that has yet another strange feature in an already strange creature. It reflects ultraviolet light; and travelers armed with such a special flashlight in the desert could watch scorpions at night glowing in fluorescent pink and green, visible from even 20 feet away. Scientists think they may have evolved this "ultraviolency" as an attractant to insects, which the scorpions can then consume more easily.

Soldiers in the Saudi desert during the Gulf War sometimes got pairs of scorpions there to start fighting, which is not hard, and not too dangerous to the observers either (the armed forces always pack antivenins for serious bites). It is still unclear, however, how some owls, bats, and snakes can actually eat various scorpion species safely.

Most scorpions are aggressive, but there are a precious few who are not. One of the African species is even social (unusual for their arachnid group). These are big mothers and fathers, up to 8 inches long, who live in mated pairs, peacefully. They tend to their offspring for two full years.

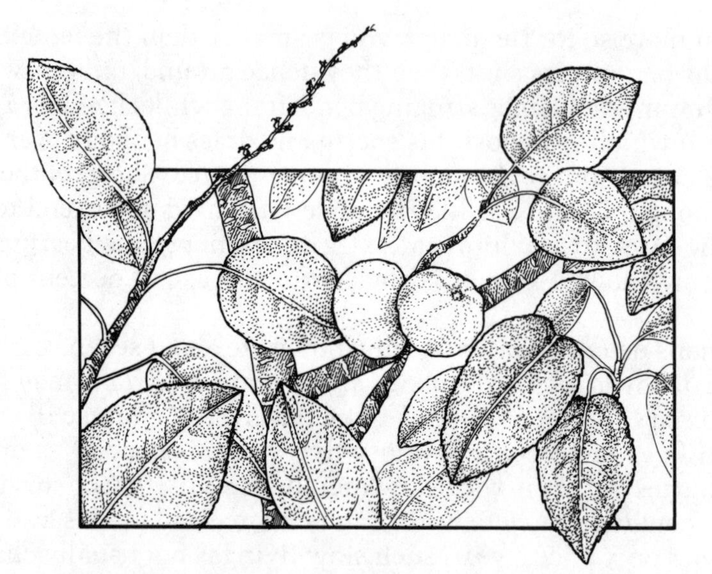

Manchineel

DANGEROUS FRUIT

Across America, there are about 20,000 species of seed plants, overall, and another 20,000-plus ornamental shrubs and agricultural plants. In this leafy array, only about 10 percent of the species are poisonous to humans—and just a few are fatally so. The manchineel is one of them.

If rain drips off the leaves of this tree onto your skin, you can get a severe rash within a half hour, then temporary blindness if you happen to rub your eyes. If the plant is part of a friendly campfire, the smoke can cause headaches and nasty eye irritation. Anyone foolish enough to actually eat its fruit or leaves will notice, one to two hours later, that lips, mouth, and throat are swelling and blistering, and that stomach pain, vomiting, bloody diarrhea, shock, and sometimes death follow.

The manchineel is a plant that grows in southern Florida's coastal hammocks, especially the Everglades and the Keys. A small sprawling tree, it has warty bark, oval leaves, yellow or red flowers, light-green to yellow fruit, and milky sap. (The latter style of sap should be avoided in any plant.) The fruit, about the size of a crab apple but sweet, has confused some people into thinking the manchineel is a guava tree. Both fruit and sap contain two forms of the toxin hippomanin, as well as two irritants (huratoxin and mancinellin). Avoid this little tree.

Nile Crocodile

SERPENTS WITH FEET

Crocodiles can stride with pride. They are part of the world's most successful group of land vertebrates: the ruling reptiles. This family dates back to the Mesozoic of 245 to 65 million years ago, and the crocodilians—the twenty-three species that include crocodiles, alligators, and gavials—are its only living representative. Their appearance has not changed much in the more than 70 million years since the end of the era of dinosaurs.

The respect one feels at this evolutionary success should be extended to influence one's own swimming habits. Stay out of the rivers, streams, ponds, lakes, and coastal areas of Africa. The Nile crocodile, which lives everywhere in and around Africa except in the Sahara, is the fiercest and most aggressive in the whole crocodilian family. Up to 18 to 21 feet long and as heavy as 2,200 pounds, it can eat anything it wants to. More bad news: entrepreneurs are now breeding a large group of Niles in Brazil, with the goal of selling their skins; and scientists there are worried that a couple of crocs will eventually escape and take over Amazonia.

One need not worry so much about the juvenile Nile crocodile. It eats only insects, fishes, and small reptiles. As it matures, it turns to larger fish and small mammals. At this point, people can still survive an attack, though they often lose a body part or two. The adult Nile crocodile feasts at the size level of the zebra, warthog, antelope, cow, dog, and human being. The way it grabs its large prey is to lie submerged and motionless in the water, often adding to its natural camouflage by staying in the reeds, under a drooping tree, or among logs or water lilies. Then it lunges suddenly with jaws agape. A crocodile is patient and usually gets plenty of food simply by waiting for it to come along. An animal or person may be swimming or wading nearby, working at the water's edge, or even minding its own business on shore. The crocodile is surprisingly fast at lunging, even onto the land, hurling itself up partly by lashing its tail and partly by levering its body to an almost vertical position. No piece of meat, including humans, is safe within 30 to 50 yards of the water's edge, over which distance the croc can even outrun a surprised person.

Nile crocodiles have been known to attack small boats too. And, once they have a hold of something or someone, it is almost impossible to make them let go. It is only a myth that making noise or staying with a group of people will scare them away. They fear very little, since the only enemies an adult Nile crocodile has are larger adult male Nile crocodiles during the mating season, or a person both armed and wise to its ways.

An apt description, penned by the intrepid Marco Polo when he saw his first crocodile in Asia, could also apply to this Nile species. They are "great serpents with feet," he wrote. Their mouths are "large enough to swallow a man whole. . . . And in short they are so fierce-looking and so hideously ugly that every man and beast must stand in fear and trembling of them." The ancient Egyptians, more accustomed to the beast and thus more respectful, even had a crocodile god in their pantheon. More Nile crocodiles are evident when the Nile floods every year, bringing the fertility to the land so important to their livelihood, even now. Throughout Egyptian history the crocodile has been both feared and honored, with people in some periods even caring for crocodiles in special pools and decorating their necks and front feet with golden jewelry.

At alligator farms such as the one in St. Augustine, Florida, tourists can view the likes of caimans (from Central and South America), false gavials (Sri Lanka), ordinary American crocodiles, Chinese alligators, Morelet's crocodiles (Central America), Cuban crocodiles, dwarf crocodiles (Africa), Siamese crocodiles, and, of course, the Nile crocodile itself. Unfortunately, Nile crocodiles have felt the pressures of an increase in human populations and, even now, a market for their skins. But they are being farmed, ranched, and studied quite thoroughly in various parts of Africa.

This is a sophisticated creature. These crocs have courtship gestures that include the rubbing of snouts and throat, while baring the throat is a submissive gesture between them. The males defend their breeding territories carefully, even occasionally killing each other in territorial battles. The females initiate mating, while not requiring fidelity of the polygamous males. These future mothers are fierce and subtle too, setting up their own hierarchies and shoving each other around to get the best nesting sites. These crocodiles are the most dangerous to people during the mating season, from November to April.

After the water matings and nest site selections, the females dig and scent-mark a burrow for their sixteen to eighty eggs (the older the crocodile, the more eggs). They cover these eggs with plenty of dirt, then lie nearby to guard them, even lying on top of the mound if there is a heavy rain. A person walking nearby who hears a loud exhaling should note that this is the mother Nile crocodile's warning signal. These patient mothers leave their nest areas only to drink a bit of water, but not to eat, for the 60 to 100 days their young take to hatch.

As the eggs hatch, parental behavior becomes even more solicitous. When the mother hears little grunts and scratches, she scrapes the dirt away and begins to take both the hatched young and the still-unhatched eggs into her capacious mouth. Proceeding to the water, she carefully washes out the hatched ones, then bites lightly at the unhatched eggs until the shells crack. Soon all the little ones are swimming around, even using their mother's back as a platform for basking in the sun. Dad appears, and both parents guard their babies for several weeks.

Once mature, crocodiles also can breathe, see, hear, and smell while almost entirely under water. They may even stay completely submerged for a half hour. All this is not for sightseeing. As Lewis Carroll wrote:

How doth the little crocodile
Improve his shining tail,
And pour the waters of the Nile
On every golden scale!

How cheerfully he seems to grin,
How neatly spreads his claws,
And welcomes little fishes in
With gently smiling jaws!

Let us hope it is always fishes.

Spitting Cobra

NOT JUST IMPOLITE

Here is what you may expect if you have been bitten by a cobra: pain; swelling in the affected area; drowsiness and weakness; an inability to talk or swallow properly because of a thickened tongue; a change in breathing; headache; blurred vision; a weak but variable pulse; falling blood pressure; slobbering; nausea, vomiting, and abdominal pain; lymph-node pain; skin discoloration in the swollen area; partial paralysis; numbness; shock; convulsions; complete body paralysis. The last stage, in which the lungs cease to function, quickly results in death.

What causes all this is the venom of the cobra, a bouillabaisse of at least thirteen different amino acids and plenty of enzymes. It is neurotoxic, meaning that it kills via the central nervous system (by paralyzing its victim) rather than via the blood and circulatory system, as hematoxic venoms do. (One of the cobra's enzyme components is under study, along with similar structures from bee venom, to see if it could lead to drugs that enhance circulation. As part of the elapid family of snakes, cobras and their kin are probably the most dangerous family on Earth—not because they are uniformly the most venomous but because they tend to live near people and in areas that lack ample medical facilities. Australia's hinterlands alone have about eighty-five different elapid snakes (though by no means all of these are lethal), and Africa and Southeast Asia are crawling with them too.

Three species of the spitting cobra live all through central Africa, and one species lives in Asia. They particularly like the savannah terrain within Africa. These cobras have a broad black band under their hoods, and some have a black underside too; others are anywhere from pinkish tan to black all over. They are usually about 5–6 feet long. The spitting cobra may not only bite but may spit first. When you disturb it, the creature rears up off the ground, spreads its neck into the distinctive cobra "hood" (with special small ribs), and lets a slurry of venom fly from holes in the front of its fangs. The snake is a sharpshooter, too, unerringly accurate at any distance under 7 or so feet. Its mark: your eyes.

By hitting the eyes of its prey, the spitting cobra is able to immobilize it with pain and temporary blindness. The crucial advice: do not scratch or even rub your eyes, though the impulse to do so will be almost overwhelming. The scratching and rubbing enables the venom to enter the bloodstream through the mucous membranes. Victims will then be likely to find the blindness permanent. Instead, it is wise to move away slowly so that the spitting cobra can't bite too, and then flush out your eyes with plenty of milk, vinegar, or water.

The king cobra is much larger than its cousin the spitter, possibly more than 18 feet long. It may actually chase people, and it packs more venom per bite—about 500 milligrams. This may lead to death in only 20 minutes—but at least the king cobra doesn't spit.

Stonefish

DEADLY AS THE COBRA

If you step on a stonefish—and its spines can even penetrate flippers, sneakers, and gloves—you will go raving mad within minutes, then die within hours. The venom from even a single one of its spines causes thrashing convulsions and screaming, followed by unconsciousness and paralysis. The poison from several spines at once is fatal unless the victim is treated very quickly. The treatment takes months and is not always successful, even if your hospital has the antidote in stock and can begin administering it immediately. The wound a stonefish makes is not hard to identify. Its center is pale, with a reddened area surrounding it. It feels hot to the touch, and the whole limb swells.

This fish looks like a stone. It remains partly buried near a coral reef or on a mudflat, waiting for food to come by. Chunky and reddish or brownish, it has a big upturned mouth into which it sucks its prey. The poison spines, erected into a fearsome armor, are used for defense. The most dangerous species live off northeastern Australia where many of the hospitals stand prepared.

Stonefish are part of a powerful family of poison packers that also includes the scorpionfish and lionfish. The latter, also known as the turkeyfish or zebrafish, is at least easier to spot than the stonefish, and is a popular aquarium fish. With their dramatic colors and stripes, stonefish are, in fact, beautiful coral reef creatures, even considering their 1-foot length and eighteen dangerous spines. Even a tiny venomous prick can cause great pain, though, as well as unconsciousness, and their venom can lead to secondary infections as extreme as gangrene. An encounter with one of these stonefish cousins can also mean permanent scars. They are, however, less often fatal than the stonefish.

The best bet is to avoid stepping on the bottom of the ocean, always wear heavy gloves down there, and don't poke your finger into a hole in the reef.

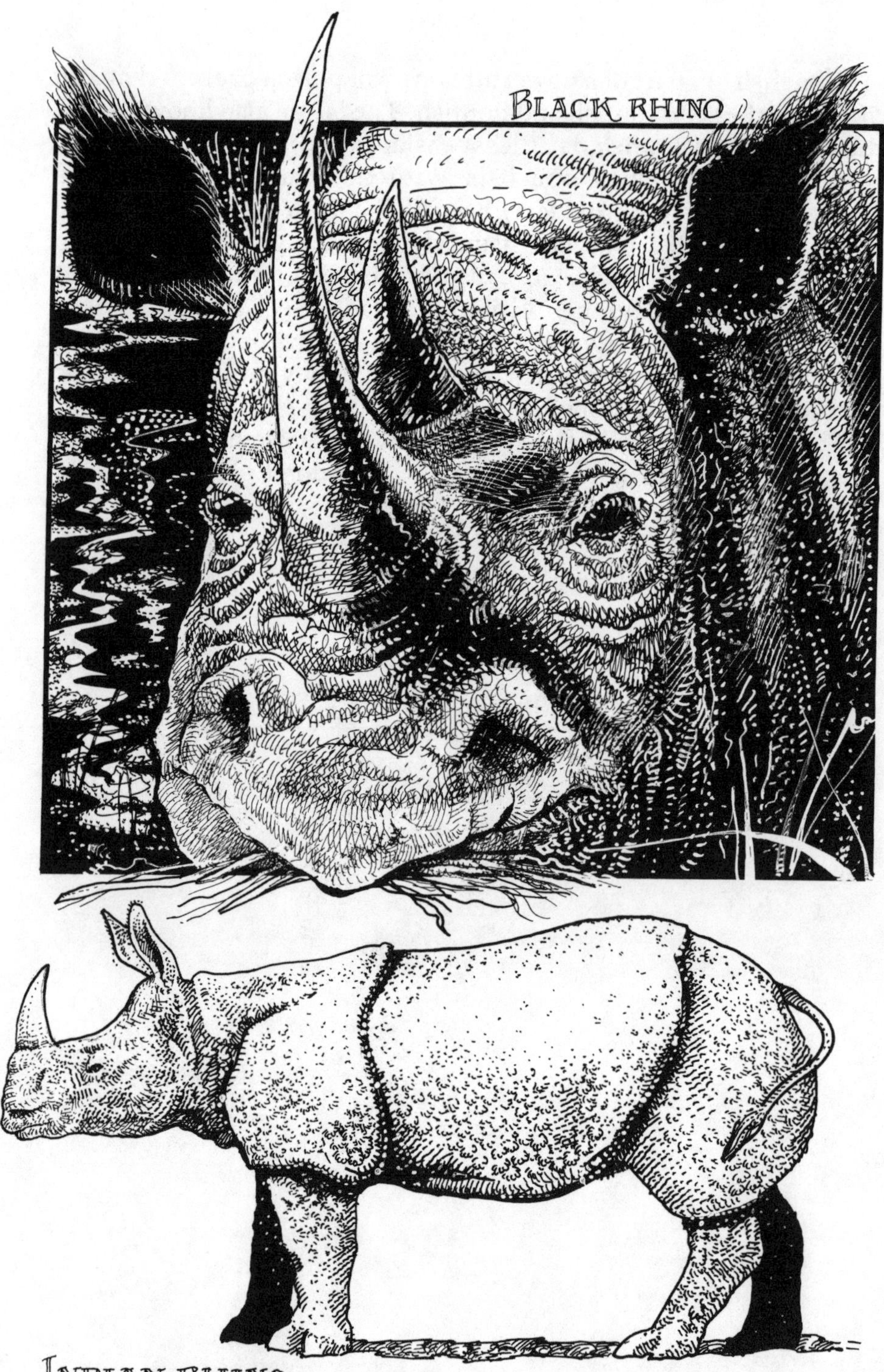
BLACK RHINO
INDIAN RHINO

Rhinoceros

ENDANGERED KILLER

There are just five species of them left: the Indian rhino, Javan rhino, Sumatran rhino (hairy), white rhino, and African black rhino. Their ancient ancestor, who lived 54 million years ago, is shared with the horse, which may well be why a rhino can run just as fast as a horse—about 35 miles per hour. As the modern rhino evolved, the creatures' snouts began to look more rhino-like and their bulk increased. One more recent progenitor, 28 feet long and 18 feet tall to its shoulder, was actually the largest land mammal ever to walk the Earth; this it did, and mightily, 35 million years ago. Later, during the Pleistocene, the vast Ice Age period of 2 million to 10,000 years ago, some of the creatures developed the now-characteristic horn, but others among the hundreds of species of fierce creatures still looked a bit like bulky horses. Their name comes from the Greek word *rhinokeros*, meaning "nose horn." Though dangerous and second only to the elephant in size among land mammals, they certainly deserve to live.

If a rhino runs after you, its horn, which can extend up to 4 feet long, is not the only threat. A rhino can trample you easily, or cut you to shreds with its teeth. You'll know when one is coming when you hear sounds like a very loud sigh, or a "phuh-phuh-phuh," or a raucous squeaking bray, or a snort, and, of course, a stupendous crashing of underbrush. Then that ugly face! They are indeed truculent, have an excellent sense of smell, and have killed people regularly. These tragedies occur most often when people surprise rhinos—they are quite nearsighted—or approach too closely when a rhino's baby is near.

The latest evidence indicates that rhinos have subtle ways of communicating with each other, not only with grunts and moans but with sounds too low to be heard by humans. Like elephants, they seem to have individual "sound signatures" in these low ranges and are able to distinguish males from females easily over considerable distances.

Sadly, not many of them are sounding or coming out of the bushes anymore. All five species are on the worldwide endangered list, with an average of only a couple thousand of each of the five remaining. Human populations push at their range continually, and poachers kill them for their horns, which may be used as dagger handles, or in powder as a supposed male aphrodisiac. Some efforts are now being made to tranquilize the animals and cut off their horns to make them less apealing to the poachers. Though this does not hurt the rhinos (the horns are made of a fingernail-like material), it interferes with their teaing at the ground to get food and with their mating displays. The horns also grow back quickly. There are now more American bisons than there are rhinos. This is a terrible denouement for a creature whose numbers in Asia alone were once a half million.

Tsetse Fly

HISTORICAL PEST

Sleeping sickness, courtesy of the tsetse fly, begins with headache, fever, and increasing lethargy. Then comes a stagger through anemia, seizures, and delirium, to, finally, coma. After that, victims just seem to sleep their way into death, a voyage that takes just a few weeks from first infection in the eastern African version of the illness, and a few years in the western and central African version.

The tsetse fly is a flying time bomb for man, woman, and beast. Across the central African continent there are twenty-two species, all armed with death and aimed at a population of about 50 million people and the cattle they need for food. In its domain, as large as the continental United States, about 10,000 people a year are killed by the tsetse.

This insect has a rich history. It may well have bitten and infected the earliest humans of all, prompting them to migrate away from their eastern African highlands. Then some one thousand years ago, Muslim missionaries climbed onto their horses and camels and set off to convert to Islam the people of equatorial Africa, a broad swatch across their continent as large as the United States. This happens to be the home—the only home—of the tsetse fly. The flies bit into both missionaries and their animals, making them drop off like flies (literally). The ill-fated expedition—along with the progress of Islam south into Africa—died of sleeping sickness. And some scientists think that the zebra evolved its stripes in connection with the tsetse fly. After all, the fly is confused by the stripes now, and zebras are invisible to them, scientists have found.

However, we shouldn't blame the fly itself for its fatal effects but rather the parasites it carries called trypanosomes, or "tryps," which are peculiarly horrifying protozoans. The "infant" insect is born clean of these creatures, from a mother fly who gives birth to a single one of these larvae every nine or ten days during her nearly six months of life. She nourishes each offspring first with milk glands inside her tiny body. The larva then burrows into the

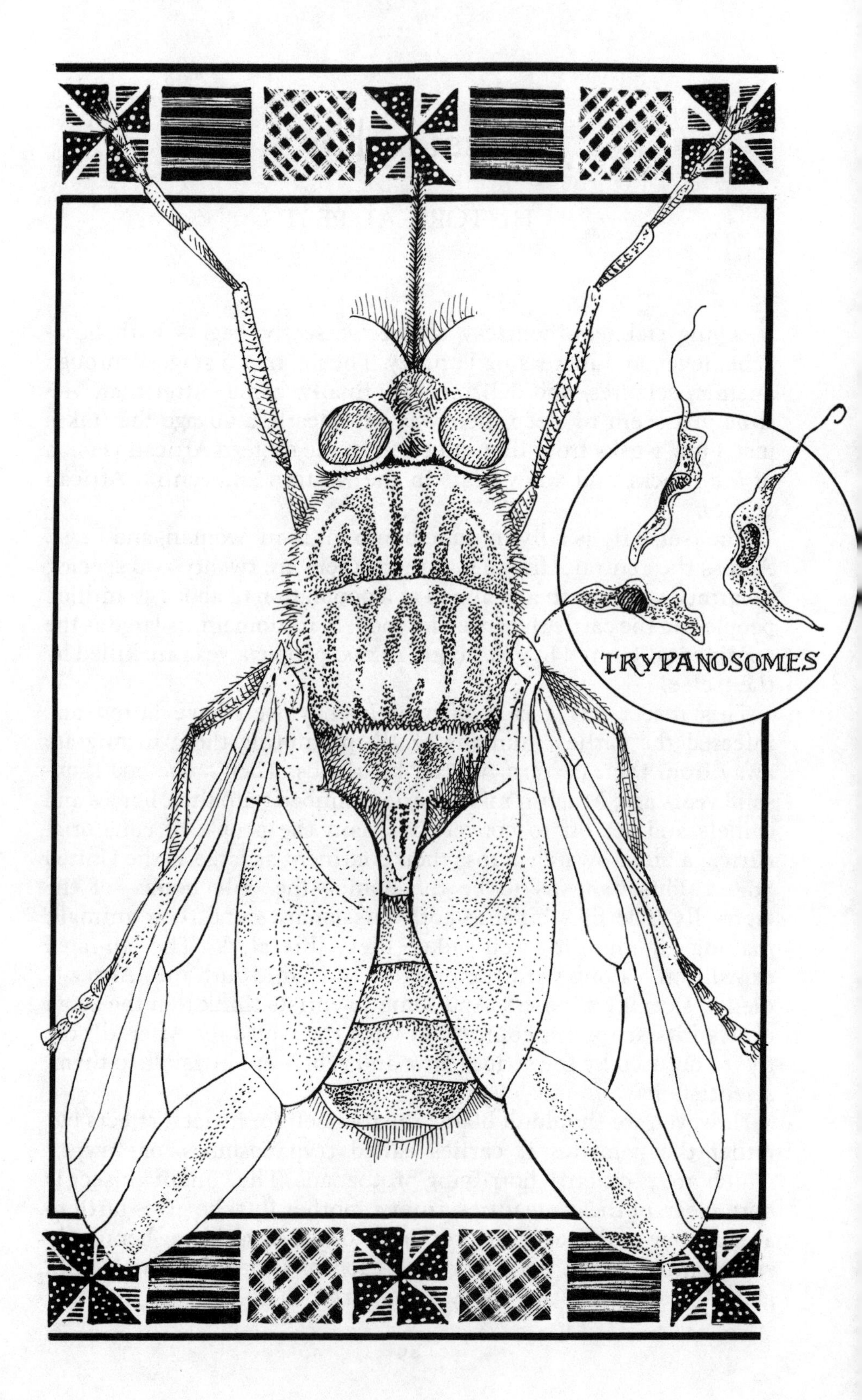
TRYPANOSOMES

ground, emerging some thirty to forty days later as a fully grown adult.

Once in flight, the brand new tsetse fly begins its search for a blood meal, digging its sharp proboscis into anything warm and alive, even through rhino hide or thick canvas human clothing, and easily through the backs of cattle. In fact, the sweet smell of cattle urine or even their breath can be used to attract the fly to traps. In attack, it will drink up to three times its weight in blood. Soon enough, its source of food will happen to be from a creature previously infected with those nasty tryps. The tryp parasites are then introduced into the young fly, to remain there for the rest of its life, turning it into a lethal weapon against both cattle and humans.

Tryps are subtle parasites, single-celled change artists. While every cell in our bodies has surface antigens to recognize intruding substances and either fight them or let them into the cell, the trick of the tryp cell is to be able to change its antigen-coating pattern approximately a thousand times. Therefore, it has been very difficult to develop a medicine to get inside these cells to destroy them. Thus, they multiply in the bloodstream and central nervous system, eventually using up the body's glucose.

The newest medicine used to fight the disease, Ornidyl, is less toxic to other body cells and seems more effective than the drug that has been used for the past forty years. But it is expensive, it must be used for two weeks at a time to combat advanced cases, and it is best administered intravenously. Not everyone can afford it or can get it, and without treatment, sleeping sickness is always fatal.

An even newer approach could lead to new vaccines. Scientists have recently learned how to disrupt the genetic functioning of a couple of the tryp species, turning them from virulent to nonvirulent. Someday, these species could be used to make vaccinations that would spur the body's own immune system. Still other scientists are working to block the tryps from being able to change their tricky antigen coats. Unfortunately, only about 20 percent of the people infected have access to any such treatments or medicines.

Another approach is, of course, prevention. Aimed at the tsetse so far have been plumes of airborne insecticides, well-placed small traps, and research into biological controls. The oddest "weapon" may be the *ndama*, an ancient strain of African cattle that seems to be resistant to the nastiness of the tsetse. More herdspeople are beginning to own them, and, as they do so, the pool of infected animals will drop. Now, herd owners are also trying to avoid uncleared areas where the tsetse—as well as precious wildlife—are still holding sway. Population pressures are making this tack more difficult, however. And the fly remains a dreadful threat.

Anaconda

400 POUNDS OF HUG

It is a myth that the anaconda crushes the bones of its victim when it kills. Actually, it suffocates the prey. The snake wraps hardest around the creature's diaphragm, then just tightens a little bit every time the victim exhales (as it must). If there is a struggle, it will only wrap harder. Soon enough, the victim can no longer breathe, the brain does not receive enough oxygen to retain consciousness, and death is imminent. Once the snake senses no heartbeat, it begins to swallow its prey whole.

This drama begins with a bite, vicious but not toxic. It sometimes ends if the prey can get away, but it is difficult to divest oneself of a several-hundred-pound snake. Anacondas surely have the advantage, since their bodies are up to 37½ feet long and 400 pounds; even one that is 20 feet long can easily kill a human being. In or out of the jungle, you are not entirely safe; special risks are encountered by circus performers who handle the snakes, and, of the few foolish enough to use this snake as a prop, several have come near death in their acts—even with plenty of help nearby.

The giant anaconda, found solely in South America, prefers to live in rivers and swamps. There, it can eat a 6-foot-long caiman anytime (though it takes a week to digest it). These snakes are strong swimmers and, if a human bather meets one in the water, he or she would not win the swimming meet. These primitive snakes have two tiny, vestigial legs, which are not retractable, but they are typical of snakes in other ways. Like all snakes, anacondas cannot close their eyes, but neither can they see well unless the prey is moving. Also, they hear through their skull bones, sensing vibrations, and they smell through their tongues (which is why the tongues flicker).

Anacondas are known for their impressive swallowing ability (as is the python, which can gulp a small impala of almost its own weight), and the key is in the design of their jaw. This they share with all snakes, yet they sport a bigger version. Snake jaws are only loosely attached to the skull bones, and they also separate into left and right sections joined only by a stretchable ligament. Throat walls are especially stretchy too, and the brain is protected by a casing of bone from being damaged during the stretch. The snake can also breathe even when its mouth is full.

Once a hunk of food gets past its mouth, the path is relatively unobstructed. The snake has no breastbone and has plenty of elastic skin. It is a creature "more subtile than any beast of the field," as the Bible puts it, even though its writers never saw an anaconda.

These snakes cannot actually swallow anything larger than about 100 pounds, though a larger adult may be dead before the snake comes up against these limitations.

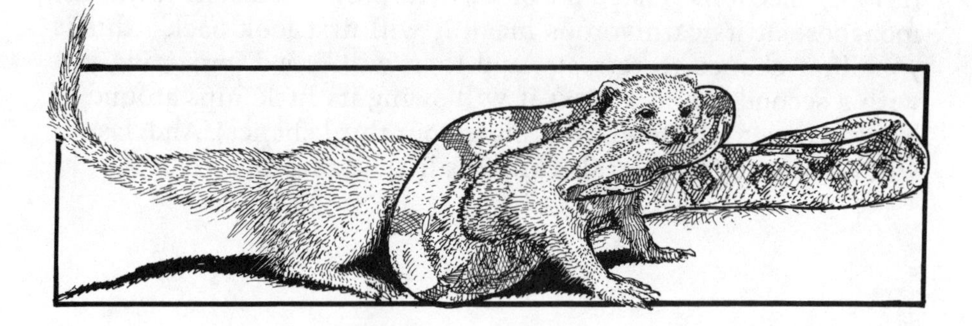

Dwarf Mongoose

DO NOT DISTURB

A mongoose is a really cute animal. It looks like a cross between a weasel and a cat, or a cross between an otter and a squirrel, depending on whom you ask. It likes to untie your shoelaces, play wrestle, ride around in your pocket, and chitter happily. Outside, it will throw eggs against rocks to open them for eating. It likes to stand up on its hind legs to look around.

You have probably heard that a mongoose can kill a snake as big as a boa constrictor, and this is true. After as much agile footwork as is required, it bites the snake behind the head and can crush its skull with its strong mongoose jaws. Thoroughgoing carnivores, mongeese also go after lizards, rodents, and whatever else is around. So they can indeed keep houses clear of snakes and mice, easily. They are also considered good "watchdogs" in general.

Of the thirty-some mongoose species, the greatest number live in Africa and are cousin to the meerkats there. Readers who are fans of Indian literature know that they live over most of India too. In the wild, the average mongoose digs its burrow in the ground or uses an abandoned termite mound, taking advantage of closable ears to keep the dirt out. The dominant couple produces young five times a year. With few enemies (mainly raptors), mongeese tend to multiply well in their warm-climate homes.

Of all the mongeese, the dwarf mongoose is probably the fiercest. It has a strong genetic tendency to bite hard, particularly to protect its food, once it has tasted the blood of its prey. If you disturb a dwarf mongoose at its carnivorous meal, it will first look back—this is your first chance to retreat—and then will growl (providing you with a second chance). Next it will swing its little hips around to block access to the food (yes, this is your third chance). And, last, it will bite savagely. Don't disturb this little creature.

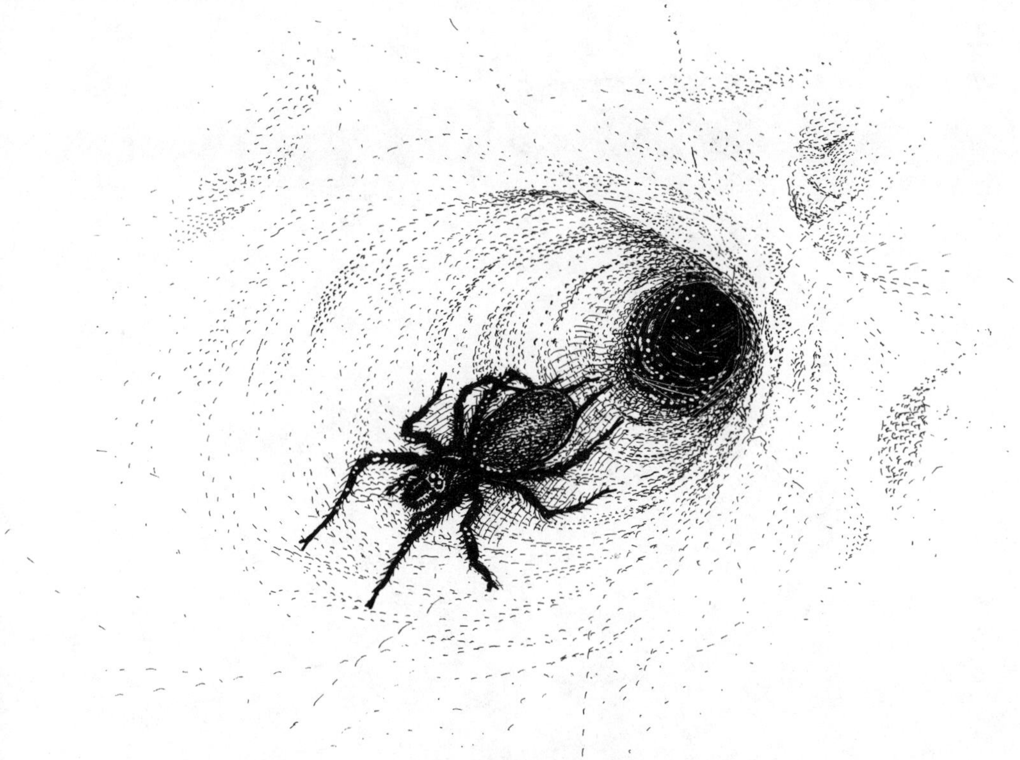

Funnel Web Spider

SMALL BUT DEADLY

All spiders are venomous, though, of course, they usually pack only the tiny doses of poison necessary to zap their little prey in the web. Quite a few spiders, however, are ample enough in weaponry to hurt us—as much as a bee sting does. And just a tiny number of arachnids can significantly hurt or even kill a person. Among them are the Black Widow (see page 167), Brazilian huntsman, aranha armedeir, bola spider, six-eyed crab spider, white-tipped spider, mouse spider, and this funnel web spider.

Of the three species of funnel web spider, all of which live in Australia, only one is truly deadly. It bites throughout northern Sydney and has killed at least several children. Another child was saved by a dose of atropine. As its name indicates, this spider is often found in a funnel-shaped web. Watch for it.

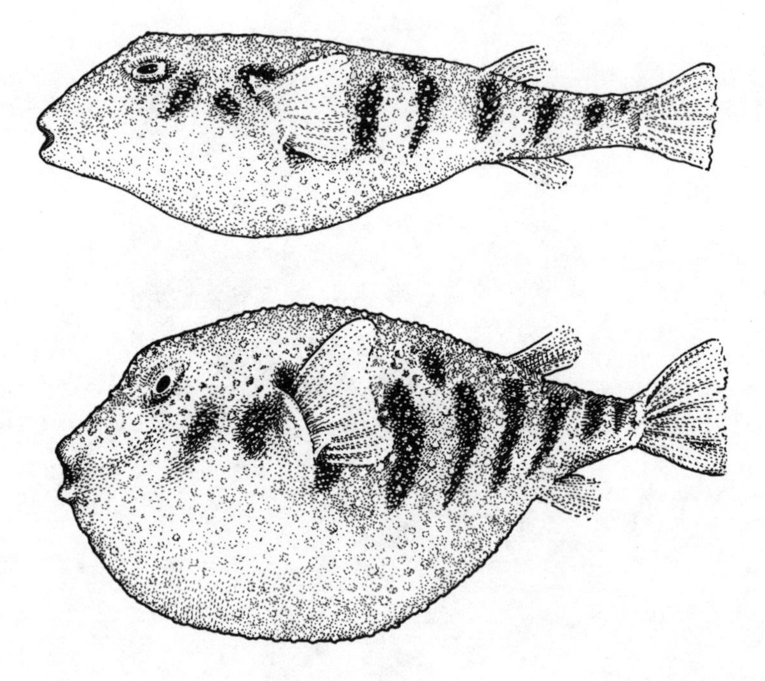

Puffer

DEADLY DELICACY

The Japanese know this fish as *fugu*. We call it puffer (or globefish, swellfish, or blowfish). The habitat in which one is most likely to encounter it is a Japanese restaurant. There, chefs specially licensed to clean and prepare it present the subtle, smooth raw flesh to diners. Carefully. The toxin found in its liver, intestines, and ovaries, which smells bad, is 25 times more powerful than the curare used on poison arrows and 275 times more deadly than cyanide. An amount that could rest on the head of a pin—about 1 milligram—is a lethal dose of this toxin called tetrodotoxin, one of the most dangerous of marine toxins. The chef's laudable goal is to make sure that the fish's organs do not touch the flesh to be served.

About a hundred different puffer species thrive in an extensive natural habitat. Puffers swim the seas near Japan and Korea, through the Indian Ocean, the South Pacific (including around Hawaii), and the waters that encircle Florida. More northerly species, those found as far north as the Carolinas, Maryland, and even Massachusetts, are not toxic. And near Japan, where their numbers are dwindling because of demand, they are being raised specially (all fully poisonous).

These creatures are large and odd. As long as 3 feet and as heavy as 30 pounds, they can, when challenged, use their pectoral muscles to suck water or even air into an internal sac. While thus inflating, they grind their teeth together noisily. All of this puffs up the fish to two or three times its normal size when relaxed. It becomes a scaly, bumpy beach ball to deter predators. After the danger has passed, the fish expels the water and shrinks. Puffers have few bones, and even their fins are not rigid enough to propel them, so they drift and flutter mildly until food comes near. Then, they quickly use their tough beak, backed by strong muscles for chomping, to tear clams, sea urchins, crabs, and others from their hard homes to become the soft meal.

The puffer's toxin is of an unusual molecular structure and creates a strange sensation. In very small doses, it can be used to treat asthma and as a pain reliever. But in the late 1700s, a man new to this fish—Captain James Cook—ate a slightly larger bit for dinner on one of his travels. As he writes in a 1774 journal entry, "About three or four o'clock in the morning we were seized with an extraordinary weakness in all our limbs attended with a numbness or sensation like to that caused by exposing one's hands or feet to the fire after having been pinched much by frost. I had almost lost the sense of feeling—[then] took a vomit." And this was only from a nibble!

Death by puffer is not pleasant. The numbness overcomes limbs, then lungs, while the mind remains lucid to ponder other choices from the menu, now never to be. Even James Bond almost found his last exploit here. Puffer poison from a sword dropped him to the "wine-red floor" in *From Russia With Love.*

How the fish itself benefits from this poison is not clear, and scientists think it may be a by-product of other processes. Puffers are more toxic to humans just prior to and at the height of their reproductive cycle, creating the suspicion that the poison is a hormone or alkaloid made by their sex glands; the females are also more toxic than the males. The fishes are also more or less poisonous depending on what they eat (more so if they have consumed some of the species of toxic jellyfish or toxic algae, for example). So some mystery remains.

More than 200 Japanese have lost their lives to this creature in the past ten years or so, most from fish cleaned at home. Many more have, of course, enjoyed the meal safely.

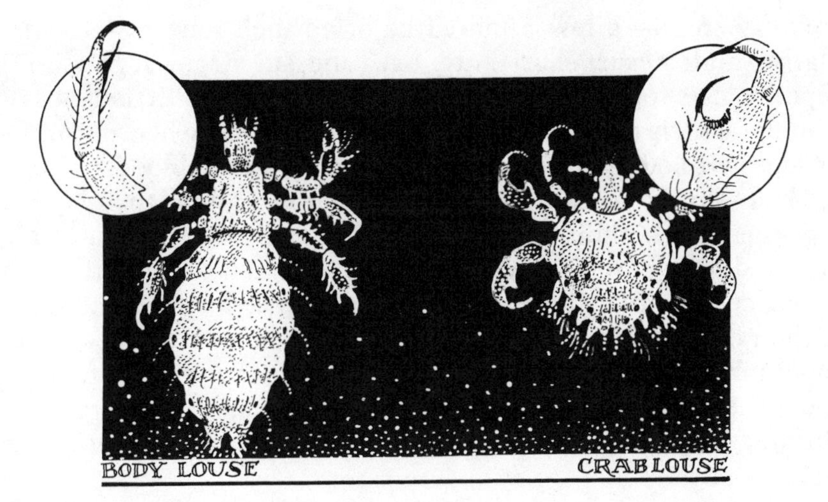

Lice

YOU DON'T WANT EVEN ONE LOUSE IN YOUR HOUSE

They have nicknames like cooties and crabs, but two species of louse, both parasitic on human beings, go by the real names of body louse and crab or pubic louse. These little lice want human blood, and they can thereby transmit typhus and relapsing fever, though mostly they just make people itch and feel very creepy. They move from one person to the other from contact with infected clothing (the body louse) or during sex (the pubic louse). Historically, they even halted the Crusades—with their lousy typhus—and also made whole Native American villages get up and move, leaving all their bedding and clothing behind. These creatures spread especially well in crowded conditions such as prisons, tenements, and school classrooms.

Our two lice species are from the order called the sucking lice, some 300 species strong worldwide and infesters of most warm-blooded creatures (save bats). Some species even live in the oceans on seals, walruses, and probably whales, where they breathe tiny air bubbles caught on the animal's skin when it submerges. All of them are very small—a few hundredths of an inch long—with particularly small heads, short antennae, and six strong legs with six superb claws for hanging on. They also have three little lances for jabbing, which they store in a little body pouch when not in use. Inside their bodies are bacteria that digest human blood.

The mommy and daddy breed right on their prey, then lay their eggs there for hatching in seven to ten days. These eggs are called *nits*, and their parents have taken care to attach them to their prey with a cement-like glue. Some species of little louse larvae get out of their eggs by taking in air through the front end of the egg, then expelling it out of their own rear ends to blast open the egg from that end. The body louse sometimes lays its eggs in bedding and clothing, but the crab or pubic louse never gets off its favorite place on its host.

Lice are terribly hard to get off—after all, they have evolved to hold on during everything from a swim to a windstorm to an athletic sexual encounter. In fact, their various adaptations can teach us quite a bit about evolution—scientists who are studying the closeness of the relationship between two species of animals sometimes look at the lice that infest them and see how similar *they* are. You will be glad to know that the pubic louse living on gorillas is very similar to the pubic louse preying on humans.

There are many more lice in the world, including about 2,800 different species of bird lice or chewing lice. More scavengers than parasites, these creatures prefer their hosts' bits of loose skin flecks, feathers, and hair, sampling blood only when the host has a convenient wound. The bark lice, also included in this group, are not even mercifully solitary and sometimes clamber over a section of tree bark like a herd of sheep.

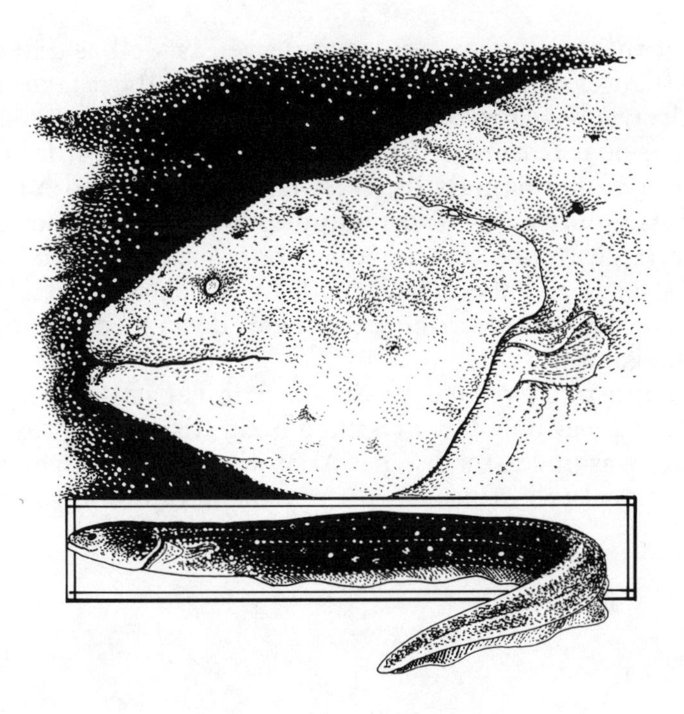

Electric Eel

SHOCKING CREATURE

This creature packs a real punch: even a baby eel can make a full-grown person tingle from its electric field, and a mature electric eel can direct up to 650 volts at something swimming nearby. Comparable to the charge of about 50 car batteries, this is enough to kill any living thing. The most dangerous species lives in the marshy but freshwater habitats of South America, growing up to 10 feet long.

The electric eel has three large electric organs that take up about three-fourths of its body. As it swims, it creates an electric field that surrounds itself and uses this field to sense prey. The eels go gradually blind as they mature and use this electricity instead of vision. Once it detects a fish, the eel directs the electric discharge right at it—zapping its dinner. To regulate the amount of charge it emits, it uses more or fewer of the backbone nerves that do the job. The largest of these eels have 200 such nerves as well as 250 cells to conjure their electricity. If the eel discharges fully, it takes a while for its electric power to build up again.

This curious creature is thick and lumpy looking. It lacks lungs and has nonfunctioning gills, so it must come up to breathe air through its mouth every fifteen minutes or so. It is also not truly an eel. The largest true eels are the moray (at about 150 different species strong) and the conger, both of which will attack if disturbed. But, lacking electricity, these eels do so by clamping down with strong jaws.

A creature more related to the electric eel is the electric ray, whose power was noted by mystified naturalists in the days before electricity was understood. What Aristotle said of it might apply to the electric eel too: it has "a sort of poison, or elixir, yet being of neither."

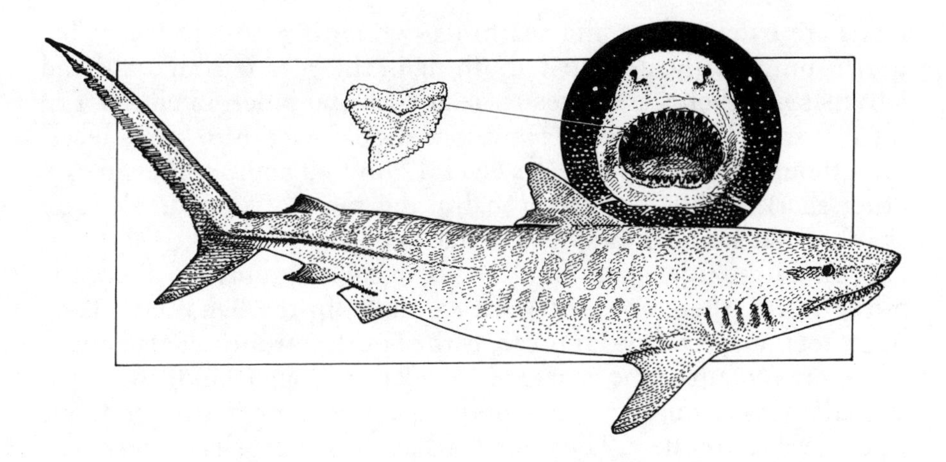

Tiger Shark

ANCIENT PREDATOR

All the sharks are ancient creatures—older than the dinosaurs—and have evolved since Devonian times about 350 million years ago. Their skeletons are knit not of bone but of cartilage; their skin is rough and elastic. Their teeth are replaceable as they wear out, enabling one shark to have tens of thousands of teeth over its lifetime. Yes, they can detect one part of blood in a million parts of water, and sense the electrical fields generated by their prey.

But few sharks are really dangerous. Worldwide, more people are killed every year by bees than by sharks. And in the United States, you run a greater danger of being hit by a falling airplane part than being killed by a shark (though all these comparisons count people who never go into the water). Sharks kill about twenty-five to thirty people per year and mutilate about seventy-five more.

The three most dangerous sharks are the bull shark, great white shark, and tiger shark, the last recognizable by the somewhat darker, but not dramatic, vertical stripes on its gray body. Its snout is blunt, its head broad, its mouth huge, its body bulky.

The omnivorous tiger shark has been nicknamed the "garbage can shark," since car license plates, logs, paint cans, and shoes have all been found in its stomach at one time or another. Its preferred foods are baby turtles and sea birds (which it grabs from just beneath), but it will eat almost anything in the ocean, from crabs and jellyfish and even sea snakes to dead dogs and fallen-in cattle. The tiger shark regularly bites Hawaiian monk seals into two pieces, even though these seals weigh several hundred pounds. It even eats other sharks. This gives it probably the most diverse diet of any shark, a diet that may surely include people.

Look for the tiger shark in all the tropical and subtropical oceans of the world (there are plenty, for example, in the Bahamas). They range regularly even as far north as the South Carolina coast and, in summertime, up to the coasts of New Jersey, Long Island, and occasionally Massachusetts. They prefer a water temperature of about 70 degrees Fahrenheit. The tiger shark is the largest shark on Pacific reefs too. Wherever they are, these creatures are 10–18 feet long, with the females bigger than the males. They usually stay in deeper water during the day, coming into the shallows to get a seabird or turtle occasionally, especially at night. They are often, in fact, in very shallow water, so certainly don't swim at night in their oceans.

Overall, they move around a lot. One, fitted (very carefully) with a radio transmitter, was observed to swim more than 50 miles in an ordinary day. They also migrate seasonally to reach their preferred water temperatures.

The teeth of the tiger shark are double trouble, equipped with sharp points for impaling prey, then serrated edges for slicing it apart. Like all the predatory sharks, they have electroreceptors in their snouts for tracking prey by the electric currents they make while swimming and, especially, thrashing.

Strangely, tiger sharks can be transformed from aggressive to actually docile when their backs are rubbed. A tiger shark whose back has been stroked a bit will stop swimming as if in a trance and gradually just sink to the bottom in a state that resembles a person in medical shock, as has been observed in aquaria and zoos when the shark was already under restraint. But I don't recommend that you give it a try! It is also said that hooking a tiger shark on a fishing line will make it much less aggressive, but I wouldn't try to get the hook out of its mouth either.

Wild Boar

WATCH FOR A GORING

Never corner a wild boar in the forest underbrush, nearby field, or bog, especially one with her striped young along. You may be charged, then gored with a pair of fearsome tusks.

Wild boars are the pigs with the most world territory. They root around from western Europe to Japan and down to Africa, and they have been introduced as wild game into Argentina, Central America, and the United States, where they have spread nicely. The moister the climate, the larger this pig will grow.

Their history in Europe is artistic as well as practical. Wild boars were painted onto cave walls by Paleolithic people, as far in the past as 34,000 to 10,000 years ago. And it is from the wild boar that the pig was domesticated about 9,000 years ago.

In Africa, they have lived with the famous too. Albert Schweitzer adopted a tame one for a while, and it followed him around, even to church. It dug its way into and out of everything—and loved to have its bristly back rubbed.

These animals are quite clever. They can recognize one another by their distinctive voices, and they have ten different sounds, each with a meaning (distress, welcome, and so on). They can be taught to come when called.

Intelligence in the animal world seems associated with omnivorousness, and these boars are indeed true omnivores. With an excellent sense of smell and their sharp tusks, they can tear up enough forest floor to make new clearings in their search for delicacies like roots, worms, and lizards. They also eat wild nuts, berries, almost any herb or grass, frogs, snakes, rabbits, baby birds, mice, even carrion. In South Asia, they have learned to open up coconuts. If wild food is scarce, these boars are happy to dig up potato, corn, and grain fields, and to grab a few chickens. They rest only at night and around noon, often in troughs they have dug into the ground and lined with grass and branches.

Wild boars are a holy terror at mating time. The males begin their fights with what is called an "intimidation march," a sort of parallel bluff and swagger. Then they bump into each other (an aggressive act among people too), try to slash each other with their tusks, and sometimes stand up on their hind legs—grunting and foaming a bit at the mouth—to try to push each other over.

They establish their territories by rubbing themselves against marking trees, also by using urine and dung as ground markers. To attract a female into its area, the male wild boar massages her with his snout, urinates, and makes some low noises. Apparently she finds this attractive.

Once the female is getting ready to deliver the young, she gathers her own mother and her grown children around her. Then on the very last day, she digs a little basin in the ground all by herself. These boars are the only ungulates who have more than two young at once. Like their parents, the baby boars wallow in the mud regularly, even in winter. They all groom each other socially too. Quite cute, if you are not in their way.

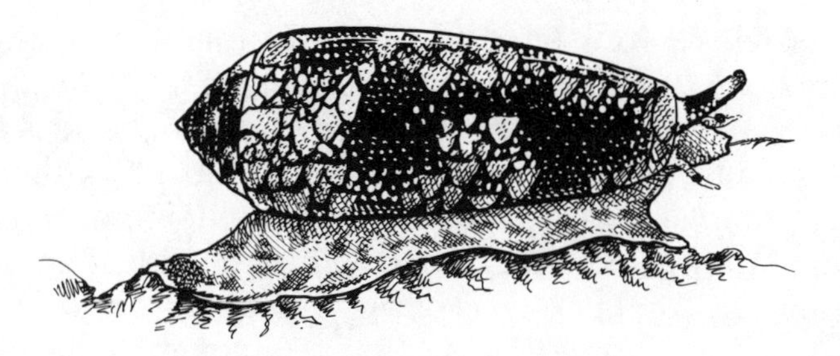

Cone Shell a.k.a. Cone Snail

PRETTY TO LOOK AT, DEADLY TO HOLD

More than 500 species of cone snails live in the sea, some grayish and brownish but some so beautiful and precious that their cone shells have been used as money in island societies. Because of their exquisite patterns, beachcombers and divers want to collect them today, too. They may not realize that these creatures are neither passive nor weak inhabitants of their lovely shells but fierce and voracious hunters who can reach out to sting, injecting so many interesting toxins into their victim that neurobiologists are only now scratching the surface of their secrets.

Scratching can be all too appropriate a word, since the toxin from only one cone makes its victim scratch incessantly, thus becoming temporarily incapacitated. This allows the toxin to take effect, creating complete paralysis. Another of their home poisons has been dubbed King Kong since it makes its lobster victims, at least, behave as aggressively as King Kong, until they too succumb. Still other cone toxins cause the cone snail's prey to become confused or even convulsive, jerking in and out of its shell until the cone can get it. And another acts as an anesthetic, just long enough to keep a fish from realizing that it has been stung and to prevent it from thrashing away.

Don't let one of these creatures get you. Some of these attractive shells have teeth inside of them a full half inch long, sharp enough to stab through cloth. The thin, barbed, hollow teeth are hidden in a venom sac behind a long proboscis, also hidden, that darts out to spear a shell collector. And all of them can easily reach up and over their backs where your hand has grabbed them. It is best not to handle any cone shell more than 2 inches long.

The first general symptoms of cone shell toxin, besides the pain, are usually a numbness and a tingling around the lips, then at least temporary paralysis of the legs and arms—and, increasingly, of the lungs. There may also be dizziness, vomiting, and, of course, difficulty in talking. Total paralysis of the diaphragm—and death—can come within a few hours. Those "lucky" enough to have been attacked by one of the less venomous species will find that these effects will actually pass, but in anywhere from a few hours up to a few days.

The cone snails whose natural food is fish are the most venomous, followed by those who feed on mollusks; the species who feed on sea worms are not fatal to people. Unfortunately, the cone snails that are the most beautiful are usually also the deadliest. Those to watch out for most closely include the tulip cone, marbled cone, skirted cone, court cone, textile cone, and geographer cone, all found at the coral reefs of the Indian Ocean, or near Polynesia, Australia, the Red Sea, or East Africa. But there are a few venomous species in the tropical Atlantic, Mediterranean, and around California and New Zealand. And, only recently, a few drab-looking but dangerous ones have been found as far north as Denmark and Scotland. Note, too, that most cones come off the reefs and into shallow water to lay their eggs at the beginning of summer.

Researchers who specialize in cone snails are amazed at the subtlety of their toxins. All are neurotoxins, but they act on different medleys of biological receptors in the prey's nerve or muscle cells. It is this specificity that makes the snails useful beyond themselves, since their toxins can be used as probes to study nerve channels. What the toxins all have in common is an ability to act fast, since these predators are, after all, snails—and can't chase fast after their food. Fortunately.

Water Hemlock

DEATH BY THE RIVER

Hippocrates wrote about poisonous plants, and so did Dioscorides, another ancient Greek, who catalogued more than 600 medicinal—and poisonous—plants. After all, even Socrates was killed by drinking one of them: poison hemlock. Water hemlock is its deadlier cousin, the most violently poisonous plant native to North America.

Who would think that a member of the parsley family—along with the gentle carrot, celery, parsnip, and others—could even poison children who make peashooters out of its hollow stems? The drastically active ingredient, found in all parts of the plant but mostly in its roots, is cicutoxin, a chemical that acts quickly on the central nervous system. Within a half hour convulsions can begin, accompanied by nausea, salivation, vomiting, diarrhea, abdominal pain, dilated pupils, fever, and delirium. The convulsive stage, next, includes a few periods of calm but then proceeds to complete paralysis, pulse both weak and fast, respiratory/circulatory failure, then death.

Plant poisons are various, it should be noted. Some, like the water hemlock's, act via the central nervous system. Others specialize in the circulatory system, and some begin by disrupting blood chemistry. Many are skin irritants, affecting the mucous membranes. Some make the skin overly sensitive to light. And some cause allergic reactions.

To beware of the water hemlock, one must be sure to identify it correctly—its death-dealing roots can easily be mistaken for those of a wild parsnip or wild carrot, and they often smell like celery. The plant grows in moist places, near a stream, swamp, or drainage ditch. It can be as tall as 8 feet high, with leaves up to 1 foot long. The narrow leaflets are toothed. In season, it is decorated with lots of small white flowers that grow in big rounded clumps like sloppy fists.

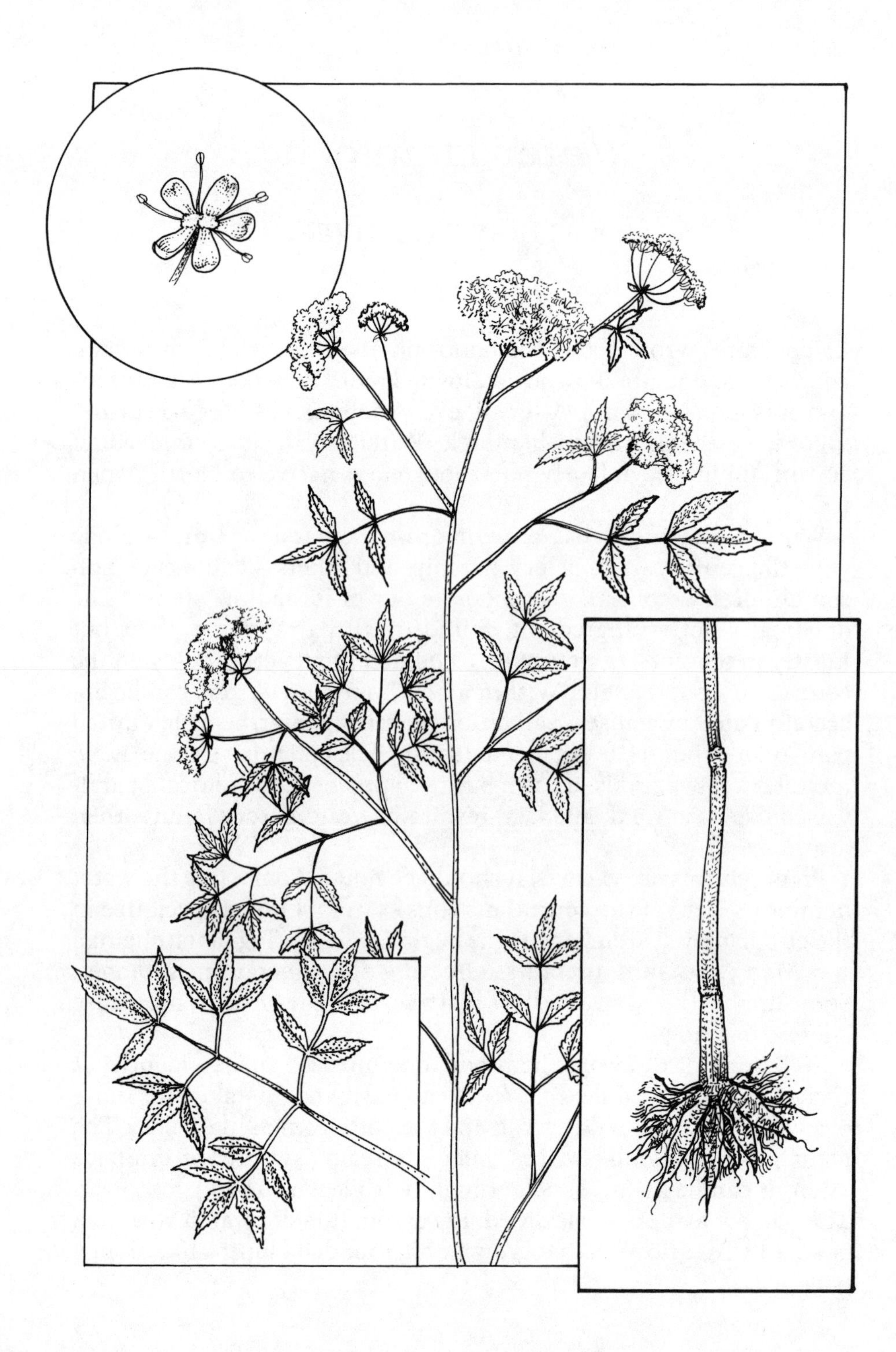

American Alligator

ATTACK EXPERT

Before the most recent Ice Age, our native alligator was even more fully native, crawling across the entire northern United States. Now, watch out for it only in the southern states, from coastal Virginia and North Carolina south to Florida, then west to the Rio Grande in Texas, and north through the Mississippi River's drainage to southern Arkansas and even parts of Oklahoma. It enjoys life in marshes, swamps, rivers, lakes, cattle water holes, and occasionally the edge of the ocean. The water can be clear or turbid.

American alligators are not nearly as dangerous as the Nile Crocodile (see page 39), but, then, they probably live a lot nearer to you. American alligators attack about twelve people per year (mostly in Florida), though they usually stay in hiding or walk right by. The attacks seem to come only when they are quite disturbed, or very hungry, or actively after your dog or cat, and almost never when they are out of the water. In fact, the attacks are often from gators who have been fed by people in the past and so have lost their fear of humans. (Some people are stupid enough to enjoy lobbing marshmallows toward them and wildlife officials in southern states have had to put up signs.) They usually go after smaller adults or children. Many of these attacks do not lead to fatalities, and the most recent American to be killed by an alligator was a little girl walking in shallow water at an inland lake in Florida in 1988.

American alligators can be content for a long time between meals and are easily exhausted after the exertion of procuring one. They prefer large fish (swallowed whole if pointed down their throats) and water birds. Their stomachs are acidic enough to digest bone well. They can breathe while their mouths are full. And, like birds, they have gizzards to help grind things up. These are quite efficient predators; in fact, their family of crocodilians is said by some to have kept almost all of the big mammal predators from evolving to colonize the sea.

The American alligator is special in that of all the loudmouths among the crocodilians it is the noisiest. These gators hiss, chumpf, growl, grunt, bellow, and roar. When one of the males

starts bellowing, the others join in, each with its individualized voice. (Their sense of hearing is excellent, their ears equipped with flaps that close when they submerge.) They will also "headslap" the water vigorously two or three times in a row.

Along with the Chinese alligator, they are the only crocodilians able to tolerate life in the temperate (as opposed to the tropical) zone. Although they prefer temperatures of 89 to 95 degrees Fahrenheit, they can survive at 39 degrees too; when cold, they go dormant but do not truly hibernate. And, when there is a drought, they are flexible, digging themselves into holes or burrows to survive quietly. They will guard these gator holes even when there is hardly any water in them.

Alligators are very careful mothers. They make a mound for their forty-five-plus eggs, guarding the little hill and then the hatchlings for several months, to keep away all the snakes, turtles, black bears, raccoons, and great blue herons that seek to eat the babies. The new, juvenile alligators are black with some yellow cross-bands, which fade by adulthood. Once grown, the adults' eyes become silvery, and their snouts and skulls become individually shaped. A male grows to about 13 feet long, with the record at 19 feet.

In the Everglades at least, these alligators can be seen basking together and are known to stay in their family groups for years at a time, the older "children" staying around as their younger siblings mature. Dominance hierarchies are also common.

Alligators get fevers when they have infections, and they have four-chambered hearts, as mammals do.

American alligator populations have gone up and down in the wild—they were an endangered species as recently as the 1960s—but there are plenty of them now. Not that this is an excuse for alligator wrestling, in which "brave" men awe tourists by appearing to defeat a gator by strength alone or to "put it to sleep." The truth is different: if one can turn an alligator over, its equilibrium becomes disturbed and its eyes cannot focus well. It will lie quietly and try to become less disoriented.

Coexistence is a better goal for the alligator-human relationship. Over the past twenty years, the human population of Florida has doubled. So has the alligator's. People who see an alligator in the mall parking lot should call the authorities to remove it to a better place. Those who observe one in the golf course water hazard should take the penalty. And humans who find one in the backyard or swimming pool should remember that a reptile that regularly grows as long as 14 or 15 feet and can weigh up to 1,000 pounds deserves respect, not the dinner leftovers. If the beast stays around, contact the Florida Game and Freshwater Fish Commission. A reptile nuisance trapper will move it for you, allowing both of you to live, separately.

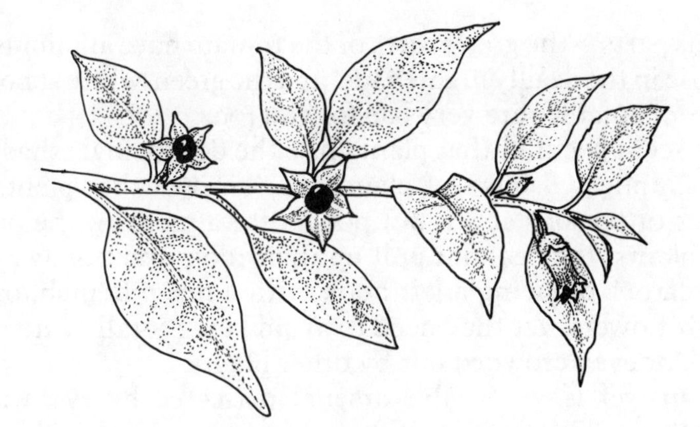

Deadly Nightshade

CHEMICAL WEAPON

Real people who wanted to be witches once used small amounts of this plant, as well as its cousins henbane and jimsonweed, to drug themselves into the sensation of flying. Off they went then on their imaginary broomsticks. Others, throughout history, have used slightly greater amounts to act out real murders.

The plant, called deadly nightshade or belladonna, is chockful of chemicals called alkaloids, not only the one preferred by the witches but another that can be used to make atropine, an eye medicine also occasionally used as an antidote to certain poisons. All dosages must be watched very carefully, since even the atropine form can cause drug dependency as well as glaucoma if it is given in excess.

The deadly nightshade is a member of the nightshade family along with a bouquet of some 2,000 other flowering plants including petunias, hot peppers, potatoes, and tomatoes. Since the dangers of the deadly nightshade have indeed long been well known, people once extended their fear to tomatoes. For a long time, and even longer in America than in Europe, no one ate them or anything made with them. It was not until the mid-nineteenth century in the United States that ketchup was considered truly safe to eat. This is not as silly as it may seem, since tomatoes and also potatoes actually have poisonous parts—the green parts of the tomato have alkaloids similar to those in the deadly nightshade, and the green-white shoots that sprout from potatoes are very bad for you too.

It may seem peculiar that plants like the deadly nightshade have such a complicated brew of chemicals, or even that plants pack medicines or poisons at all. But ponder for a moment the predicament of plants: they cannot pull up their roots and run away from their predators, or swing a left hook with a mean branch, or sting with their flowers. Yet they need to avoid being totally eaten up, or trampled, or even crowded out by other plants.

Their answer is a veritable arsenal perfected by evolution for more than 200 million years. It includes, across the plant kingdom, thorns and prickles and sticky resins. A few plants use camouflage (just try to find the cactus called flowering stone) and mutualism (harboring ants or other stinging insects to help out). But chemical poison is the specialty of plants. They make almost endless insecticides, fungicides, antibiotics (to fight off invading bacteria), and toxic substances. The alkaloids found in many plants, including the deadly nightshade, interfere with the nervous systems of many insects who are after the plants. More than a hundred fungicides have been discovered in plants so far. Antibiotics are common in sycamores, wheat, and many other plants too. Plenty of toxins ward off rapacious birds. Still other chemicals, such as cinnamic acid, are seeped into the ground by a tree like the guayule to keep other plants from growing close to it. Some of those plant poisons are quite accidental by-products of plant development that have turned out to work well.

Plant chemicals, whether in the form of alkaloids or a brew that includes the likes of glycosides, oxalates, phytotoxins, minerals, or even polypeptides and photosensitizing compounds, can usually be produced by the green warrior in greater amounts to greet larger infestations, especially of insects. Sometimes plants are even able to signal nearby fellows to produce the chemicals before the insects spread to them next. Not all this nasty chemistry affects humans, of course. We can enjoy citrus peel, which is fatal to many insects. By the same token, the old wilderness advice to judge berries safe if birds are eating them is false. The bird may not go into convulsions on the hiking path, but the hiker may.

This brings us back to the berries of the deadly nightshade. Lustrous black or purple-black and up to a half inch across, they are the most dangerous part of the plant and can kill quickly. Just three berries can kill a child. The flowers on this deadly bush or small tree are also very toxic—look for them to be bell-shaped, up to an inch long, and purplish brown, dull red, or greenish yellow. The leaves, crowded on the short branches, are poisonous as well. And digging up the roots would be foolish.

Symptoms of this nightshade's poison are beginning when your mouth feels dry and it becomes hard to swallow. Next comes flushed skin, a quickening heartbeat, dilated pupils, then increasingly blurred vision. After that, blood pressure rises, the pulse feels strange, and feelings of excitement, delirium, and confusion mingle. It becomes impossible to urinate. Then comes coma (in the context of a below-normal temperature), followed by respiratory failure, and death. The deadly nightshade's name comes from the tranquil feeling that you experience as you slip into the coma.

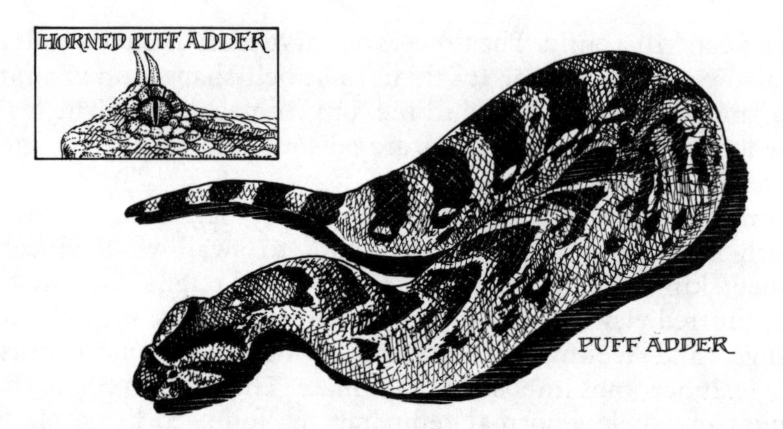

Puff Adder

FEARSOME SIGHT

This snake is one of the gaboon vipers, the family with the longest fangs in snakedom. These teeth, which can extend to 2 inches long, are able to penetrate tough shoes and thick clothing. And, if they belong to one of the larger of the puff adders, they can pack enough venom to kill a person five times over. Overkill from the underbrush, indeed.

Puff adders are common across tropical and southern Africa, and can occur in the Arabian peninsula too; they prefer savannah and forest but avoid only the rainforest and the emptiest desert terrains. Because of their nasty disposition and effective camouflage, these adders probably kill more people than any snake in Africa. Anyone who steps near one will see a fearsome sight—a snake up to 4 feet long that is puffing itself up and hissing loudly before it bites.

The puff adders are particularly chunky snakes, with large heads and ample venom glands. As in all snakes, these glands (a specialized salivary gland) are vaguely triangular and lie behind or below the eye, connected to the upper lip. The puff adder's eyes have vertical pupils and clear shields to cover them when the snake buries itself in sand or dirt. One species in the family, the horned puff adder, also has a little pair of horns and comes in all desert-y colors.

Listen to this cautionary tale told in parts of Africa: the puff adder supposedly has a magical ball of grease in its tail that makes it detest human excrement, and if you touch it with a stick dipped into this substance, it will pursue you to the death. All over the world one can hear fear stories like this one about snakes. One reason surely is that dangerous snakes are gliding along every continent except Antarctica. In much of Asia and Africa, they kill about 5 people per 100,000 every year. One province of Burma holds the world's record: 36.8 deaths per 100,000 people annually. Australia also has plenty of deadly snakes. And venomous snakes are found, in smaller number, all over Europe as far north as the Arctic Circle.

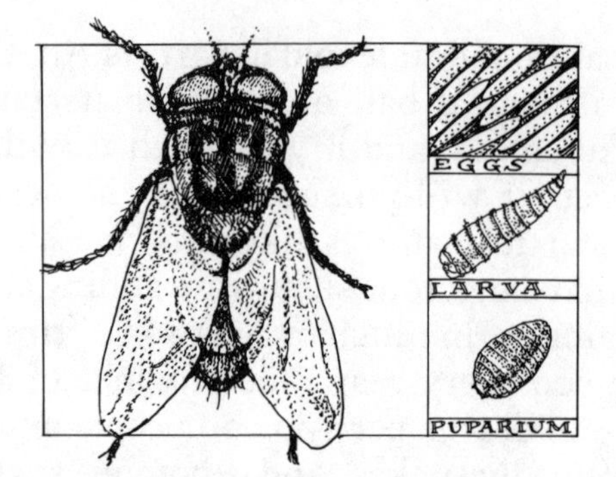

Screwworm Fly

AND SOME RELATIVES

The habits of the screwworm fly make even blowflies and fleshflies seem charming by comparison. The latter two species hurry onto newly dead animals to lay their eggs, which then hatch into maggots. These maggots proceed to use their own excrement to soften the dead meat. Then they eat it.

Screwworm flies perform a similar trick—but on animals and people who are very much alive. They aim straight for a wound, even one as small as a tick bite. The female screwworm fly lays in it, 500 to 3,000 tiny eggs over a three- to five-day period. The eggs hatch in about twenty-four hours, and the new maggots then feed on the wound, next moving into the body to eat more. Cattle can be killed in a few days to a week; this happens once the fly enters the brain. In people it is similar, with death coming when the maggots, each about half an inch long, reach the brain or lungs.

After a week of such deadly feasting, the maggots are mature enough to leave their host. They fall to the ground and pupate, emerging soon as adult screwworm flies, ready to start another disgusting little life.

Screwworm flies are found across South America and the Caribbean, parts of North America (including the southern United States), and are now entering Africa, after a boat ride over on infected sheep. In a new government program, 40 million sterile male screwworms have been sent to Libya, where they outnumber fertile males by about 10 to 1. The idea is for them to mate with the local lady screwworms, result in zero babies, and thus prevent the fly from spreading all over Africa from its Libyan entry point. With about two decades of similar projects using sterile males, the screwworm has been pretty well eradicated from the western United States. Without such efforts, their damage in a given area can, as one scientist said, "make locusts look like nothing."

The screwworm fly has some pretty odd relatives too. Some eat nectar and pollen peacefully. And some don't harm us even if they are a bit pesky. Others are so desirous of carrion that they are fooled enough to focus on the just-as-smelly skunk cabbage and the stinkhorn fungus, thus pollinating these stinkers instead of finding the carrion.

The housefly, much more a part of our lives than the worse screwworm (and not closely related to it), can be pretty horrifying itself. Its favorite attractors are garbage, excrement, and dead animals. But it will lay its eggs, twenty-five to a hundred of them at a time, on anything that is warm and rotting. These eggs hatch in twelve to twenty-four hours, whereupon the maggots remain in place for a couple of weeks. It is said that if all the eggs from one mother housefly lived, she would produce more than 5 trillion offspring in just one season at the latitudes of Washington, D.C. Later, as adult flies, these creatures move from garbage, excrement, and such to our sugar bowls and faces, where they spit up material from their last meal to soften this one. Their mopping and sucking mouth parts make them the most highly evolved of flies, here since the days when flowers appeared. One fly, dissected after a meal, was found to contain a full 6 million bacteria in its little body. At least these weren't screwworm flies.

Black Bear

LESS DANGEROUS THAN YOU THINK

Although they are often thought to be dangerous, about the worst thing a black bear might do is chase you for a while if you surprise it in its territory. It may also bluff, that is, charge at you but stop before coming too close. Just stand your ground or shout at it and you will scare it away! Usually, these bears just stare at you or climb up a tree to get out of your way. In fact, they try their hardest to keep far from people. And only the female is territorial (though the males mark trees during mating season).

One Canadian naturalist I know was once chased—he admits he was a little scared, in spite of what he knows about black bears—and he ran away fast. He knew, too, that these bears can run at 25 miles per hour top speed, and that he could not. He was also aware that an adult bear can weigh up to 550 pounds. But within a few yards he no longer heard the crunch of sticks and the scuffle of grasses behind him—and he turned around. There was the bear ambling in the other direction, its don't-give-a-damn and I-scared-him attitude evident in the swagger of its rear.

Black bears have killed very few people, and almost entirely in self-defense. One American bear expert puts the number at 25 people in the entire twentieth century, as compared to about 70 people killed by dogs, 160 by tornadoes, 180 by bees, and nearly 200,000 by cars *annually*.

This animal is not very aggressive, even when its cubs are along. Why? Because all of its life it has found safety, food (acorns and fruit, for example), and contentment by just climbing up a tree. Black bears are extraordinarily fast climbers, and up the cubs go with the mother even on the day they leave the den. In fact, a mother bear often uses trees such as the white pine as babysitters, leaving the cubs at the base, since she knows how quickly they can climb them.

Nonetheless, it may be useful to know what black bears mean when they make certain sounds and exhibit certain postures. A grunt is a friendly, social sound, and a grunting bear will never even bluff an attack. Blowing, with upper lip flared out, means mild-to-medium fear or threat. The gargle, especially with bared teeth and ears pressed back, is a threat gesture: get out of its way. Threats are just threats too, almost never preludes to attack. Also, a fast "oh-oh-oh" noise means pleasure. And a bawling sound means pain. When bears are happily playing, they make no sounds at all. If you see a black bear high up on its hind legs, that just means it is trying to see, hear, or smell better. And if a bear enters your tent, it is because your perfumed deodorant, shampoo, or insect repellent was so strong that it masked your human odor. Just wave your arms and yell at it to leave.

They are solitary animals and there are probably not many around in a given area. Black bears can be chased away from camp sites easily by people and are very afraid of fires (as forest animals, they have honed that fear well); they have sharp fangs but use them mostly to rip open old logs, to eat the insects (especially ants) inside. This animal eats a few baby deer, rabbits, and birds and some carrion, but mostly enjoys fruit, nuts, and vegetables such as pussy willows; it will invade backpacks, garbage cans and dumps, but only when especially hungry. In spring and summer their search for food takes them mostly up trees, in autumn they are looking at the forest's edges, and on very hot days they are either hiding in the shade or going for a brief swim. In the winter they are in their dens.

These animals are generally clever. On one cross-species intelligence test (a dicey business, to be sure), they scored higher than dogs. Their hearing is about twice as good as ours. They know how to sneak away quietly when their predators—people—come near. Their smell is so good that when they sniff a log, they know whether there are ant pupae (a delicacy) inside. Their vision is also excellent. They have claws designed for climbing and digging, even hollowing out a big, snug den with leaves, grass, and twigs added for a bed. Male and juvenile bears den alone, mothers with their cubs.

In their hibernation, black bears are truly extraordinary. For the whole winter they don't urinate, defecate, drink, or eat, yet they manage to survive. What their bodies do is shut down the kidneys and recycle the urine into nitrogen, then use it along with their fat reserves to build protein on which to live—all without clogging their arteries. Their hibernating metabolic rate is about half of their regular rate, and their heart rate slows. This combination of bodily magic is practiced only by black bears, polar bears, and several other bears.

But if you are going into the woods and are still afraid of black bears, take along a squirt gun filled with cayenne pepper and water, which mail carriers often use for dogs. Squirt the solution into the bear's eyes. That will make it stop to rub its itchy eyes but will not hurt the animal. I hope you won't use it, though, on these big, chubby, furry woodland friends!

Note: in the account above *grizzly bear* cannot be substituted for *black bear.* A black bear is black or dark brown or cinnamon-colored or even blond (with individual bears sometimes changing as they get a new coat every year). A grizzly is usually medium brown and always has a hump between its shoulders. Unlike its black cousin, it is truly dangerous.

Salmonella

A BIG BACTERIA FAMILY

This family of about 1,500 different species of bacteria has nothing to do with salmon (the fish) but rather with Daniel Salmon (the pathologist who first identified them). Salmonella have their ways of getting into the intestines of people or other animals, to become parasites within the cells. They are deadly enough that they are included in treaties banning biological warfare. About 500 people die every year in the United States from the salmonella bacteria, and another 35,000 or so suffer mightily, then recover. Although the role of salmonella in food poisoning is the most important to us, it is worth noting that other members of this family cause stillbirths in sheep, give fatal diarrhea to pigs and cows, cause typhoid fever, and much more.

While you have to ingest several tens of billions of these salmonella bacteria to get sick, they are certainly around in great numbers. Tiny, rod-shaped, and, of course, invisible, these intestinal bacteria often begin their job of infecting humans even in the pediatric ward, then can attack us later in restaurants and other public eating places, and for the rest of our lives. We can get salmonellosis by eating pork, beef, poultry, or eggs that have not been cooked enough to kill these bacteria. Another way to get the disease is from anything—though usually food, or water, or the equipment that is used with them—that has been contaminated by an infected person's or animal's feces. Even a person who had salmonellosis years before can still have enough of the bacteria in the feces to infect someone else.

Salmonella are a major health problem in the United States and elsewhere. Some even say their danger is increasing, since the animals we eat have been given so many antibiotics that any salmonella they have is, by definition, resistant to most of the antibiotics that we would need to take as medicines against it. Also, as more people eat outside of their homes more often, the danger increases. Hospitals are hardly immune to outbreaks, and a major one occurred in 1987 in a New York City hospital. The culprit was raw eggs used in mayonnaise made on the premises.

Fulmar

SEABIRD WITH A RUDE HABIT

This seabird looks like a sea gull but has a bit stubbier bill and seems a bit plumper. It is 18 to 20 inches long, with a wing span twice that. The fulmar's flight is somewhat different from a gull's too, more of a flap than glide, and up into high arcs. Look for yellow at the tip of its bill and those dark brown eyes. All this is important because a sea gull won't spit up on you—and a fulmar will.

Out of about 100 billion birds in the world, the most common are the shearwater species, which includes the fulmar. Charles Darwin believed that fulmars were indeed the most common bird of all. You can see them literally by the millions in the North Pacific and North Atlantic regions. They breed from the high Arctic to the sub-Arctic, and even as far south as the north coast of France. About 2 million come to Iceland alone in summertime, to nest on cliff ledges by the sea or as far as 30 miles inland. They are gregarious. And their numbers are increasing all the time. (There is an Antarctic fulmar, too, but the two species never meet.)

Halfway in from the fulmar's bill tip to its eyes are two small holes. If you are close enough to see these holes, or even the bill clearly, move before the bird goes into action. It will spit up on you in defense! Out of either of the holes, or the bill itself, or both—careful observers have differed here—will come two narrow streams of "vomit," which can land on you from about 6 to 8 feet away. The liquid is not really vomit, but an oily distillate of one of the fulmar's recent meals of fish. If hit, you will smell like old, dead fish for quite a while. A fulmar is most likely to spit up on you when it is trapped somehow, or when it is nesting and cannot really fly away. I once walked to within 10 feet of a nesting fulmar on an isolated Iceland island. Though I watched carefully and did not approach closer, a friend who did so just escaped the bird's nasty fish effluent.

Once born and fledged, the baby birds spend several years at sea before looking for a colony in which to nest. During this time, they will wander as far south as Japan, even Baja California. People ate fulmars in the past, but it is said that they don't taste very good. Somehow this doesn't seem surprising.

Eastern Cougar

PREDATOR ON THE REBOUND?

Teddy Roosevelt called it that "big horse-killing cat, destroyer of the deer and lord of stealthy murder with a heart craven and cruel." This is a bit excessive, but the cougar is indeed a paragon among predators. It can swat a small animal down with one flip of its retractable claws, or sneak up to within a few feet then pounce down with all its strength onto the back of the prey. A cougar can even kill prey as large as an elk, grabbing the elk's head between its paws and twisting its neck.

A cougar could indeed kill a person, and easily, but there do not seem to be enough of them around to fear. There have been, in the last few years, however, quite a few cougar sightings, mostly in the mountains or forests of Maine, Vermont, Connecticut, and Pennsylvania, but also in all of the eastern states. Nobody has managed to get a photograph, but some of these sightings are by wildlife experts who presumably know a cougar from somebody's kitty. The eastern cougar once roamed the eastern United States, then was thought extinct, but it may have merely retreated into Canada for a while, returning once it noticed no one shooting. They are now protected. People who have seen one usually feel somehow honored, somehow reassured that its presence means that wilderness remains.

Eastern cougars are about 5 feet long, including the long tail, weigh 60 to 125 pounds, and are usually tawny, sometimes dark brown or black.

The nomenclature is a bit confusing since people call them cougars, mountain lions, pumas, and panthers interchangeably. All the names refer essentially to the same animal, with the western cougar subspecies being a bit larger; and, with more empty land out West, their populations are greater. The American Panther (see page 211), another subspecies, is doing very poorly. Overall, the species has been able to adapt to habitats from the Everglades to the desert, mountain to jungle, but it does need space away from people.

This is an elusive creature, shy of human encounters and afraid of dogs' barking, and so very rarely dangerous. Solitary except in mating season, it hunts through a broad but definite territory. Contrary to popular stories, it does not scream but makes only quiet whistling sounds, with a few yowls during mating time.

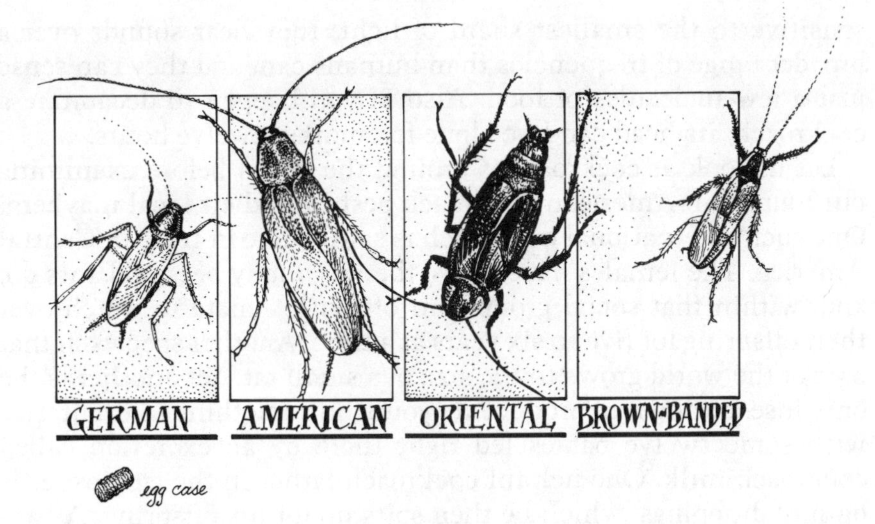

Cockroach

THE UBIQUITOUS

Our world is crawling and flying with more than 800,000 insect species, which thus outnumber all other animal species by more than two-to-one. This is to say nothing of insect individuals! Cockroaches represent only a small chunk of these—about 55 species in North America and at least 4,000 species in the tropics. But they manage to make trouble, nonetheless, not by biting people but by spreading various diseases and by causing allergic reactions, primarily through their skin flecks and feces. These feces are light enough to become airborne, then inhaled, and invisible enough to be eaten when they have fallen onto food and utensils.

Not only disgusting, they are also interesting. Few people probably appreciate the fact that cockroaches are smart enough to solve a maze (after some practice), equipped with taste organs in their little mouths and also gizzards to crush and grind up their food, able to synthesize their own vitamin C, and (like some other insects) boast sixty tiny tubes to produce their feces, which they then excrete in tiny sausage shapes. They also can navigate backward, they have big compound eyes with a crystal reflecting layer that makes them sensitive to the smallest shard of light, they hear sounds over a broader range of frequencies than humans can, and they can sense just a few molecules of food. Also, if you happen to decapitate a cockroach, its head can live alone for another twelve hours.

Let us look at cockroaches around the world before examining our four main American cockroach pests and their local mayhem. One such international cockroach is said to live in trees in Central America. The females have nests, the way many bees and ants do, and, within that social group, pair off with a mate to watch over their offspring for five to six years at a time. Another species in that part of the world grows to the size of a small rat. Yet another is the only insect known to produce its young in something like a uterus, with some twelve babies fed right there by an excretion called cockroach milk. One helpful cockroach father in the tropics feeds on bird droppings, which he then spits up for his offspring. A Panamanian species (which grows to several inches in length) lives in bat caves, and the largest cockroach of all, at a bit over 6 inches, lives on a plateau in Queensland, Australia. Perhaps the most famous of the tropical species, though, is the Madagascar hisser, just 3 inches long; these roaches hiss and butt heads when they fight. Smaller cockroach species are crawling all over the planet too; they come in lovely colors, from red to green to blue with stripes and dots. One tiny species follows leaf-cutter ants into the underground fungus gardens they maintain, feeding there.

The ancestor of all these marvels is the 6-inch cockroach of the Carboniferous era; its fossils date back to 280 to 400 million years ago. Roaches are thus as old as the huge dragonflies with wingspans of 2 feet and the ancient mayflies of the Carboniferous swamps. Some scientists think the roach has changed the least of any living thing across all these eons. It is a superbly evolved and versatile creature, and it has persisted not by finding one tiny, perfect, ecological niche but by being *un*specialized.

The four pestiest cockroach species living in the United States should inspire respect (along with aversion). They are, after all, commensal—eating at the same table—with humans. The most common four are the German cockroach, the American cockroach, the Oriental cockroach, and the brown-banded cockroach. The German species breeds the whole year round, protectively carrying its thirty to forty-eight eggs for three whole weeks, all wrapped up in a capsule; such care and fecundity could pay off to the tune of 30,000 offspring in one year alone if all children and later descendants of the mother were to survive. The American cockroach, larger at about 2 inches long, prefers the dark, but it often moves readily from apartment building to apartment building as extermination efforts progress. The Oriental roach has earned the nickname "waterbug" because it hangs around water pipes and sewers. And the brown-banded roach, the smallest of the four, particularly likes warm and dry areas; this is the species known for living inside everything from radios to alarm clocks.

All these household friends are omnivores. They will eat anything that is of even partly organic origin: shoes, vaseline, book bindings, feathers, natural-fiber clothing, camera film, human urine, paper, soap, ink, as well as every imaginable kind of human food—and most of the containers it comes in. They are patient too, capable of surviving for about three months without food and thirty days without water, and are equipped for enduring adverse temperatures and even radiation in much higher doses than humans can. No puny ecological niche for this creature.

To get their dinners, cockroaches employ caution. Nocturnal, they scurry away when confronted with light and are leery of sounds (which denote predators to them). They are especially sensitive to tiny movements of air anywhere around them; adult cockroaches have about 220 tiny hairs on each of two rear appendages that bend with the flow of air, signaling the legs to run away. They are among the insect world's fastest runners.

About 10 to 15 million Americans are allergic to cockroaches, and not all of them realize it. They may get a runny nose or a rash, or an asthma attack, or may even suffer life-threatening shock after enough exposure to those ubiquitous feces and skin bits. Energy-efficient, airtight homes have only increased this problem, since cockroaches dislike ventilation (even though they don't have to contribute to the heating and air-conditioning bill).

Cockroaches also dislike diatomaceous earth (a white, dusty powder that is actually composed of tiny fossil plant parts) and white boric acid, army ants (they will flee from them, if you happen to have any around), and a beer-soaked paper towel—they like the beer, of course, but it can make them drunk enough to fall asleep, and then you can drown them or squash them. Some insecticides seem to be effective in eliminating them.

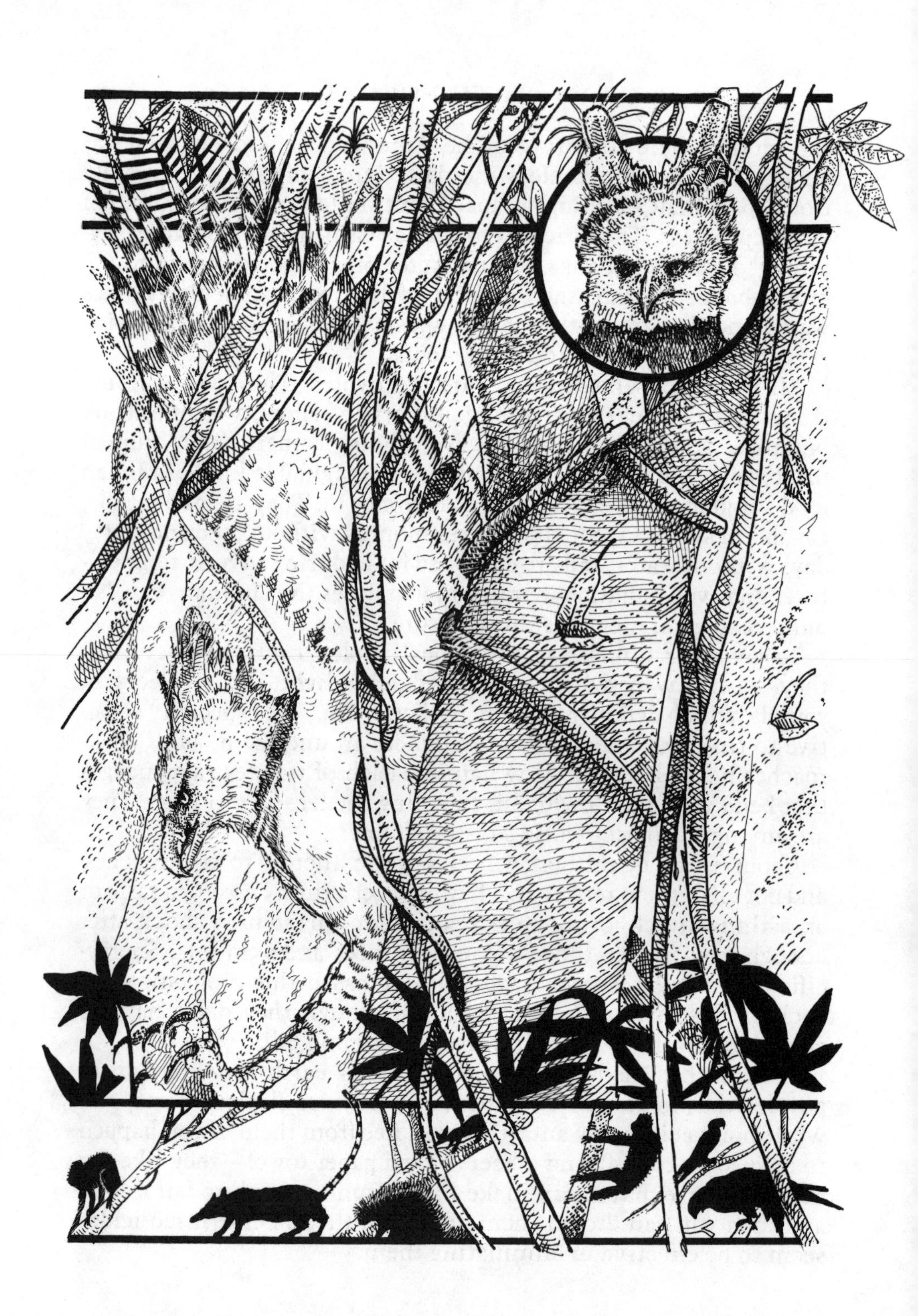

Harpy Eagle

A DIET OF MONKEYS

All eagles at ease do rollovers in flight, free fall for fun, then dive and rise on plumes of air. Then some of them get down to business and grab a monkey for dinner. This is the food of choice for the harpy eagle, the largest and most powerful raptor in the world. Its claws are the strongest of all the eagles, fierce yellow things trimmed in black. Its brown and white feathers and black face even make it look a bit like a monkey. The harpy is named for the harpies of Greek mythology, creatures with the heads of women but the claws, wings, legs, and tails of birds.

Fond parents of small babies who find themselves camping in the jungles of the Philippines or New Guinea or in the rainforests of central and northern South America would do well to keep any monkey-size offspring safe inside the tent. The harpy, at more than 15 pounds, and with a wingspan of 6 feet, can fly through thick forest at 40 to 50 miles per hour. It demonstrates hunger by fluffing up its neck and head feathers.

The harpy eagle is endangered, with probably only a few hundred harpies left, all with impressively crested heads and an ability to lift prey as heavy as themselves. They pounce on it from the highest points in the forest canopy, where they nest in the tallest trees, after detecting movement through the foliage. They eat not only monkeys but large birds such as macaws, and even sloths. The remains of these victims reappear in the pellets regurgitated out of a harpy's mouth after a meal. The waste bones and fur of the prey are thus eliminated, as with all birds of prey, wrapped up together in little oblong packages.

Eagles in general are excellent hunters, with some species stealing sheep and even small antelopes in their talons, others zeroing in with their superlative eyesight on small, hard-to-spot rodents, still others safely killing poisonous snakes, and some picking up slippery otters (some of which have been seen thrashing enough to get away, however). To find all this food, the eagles of the world fly superlatively well, their wing surface areas even larger for their body weight than most birds. They specialize in soaring. And of course their eyesight is excellent in flight. In fact, if our eyes were proportionally as large as theirs for our bodies, each human eye would weigh several pounds. Most eagles have a much better reputation than the harpy. Not only has our country chosen an eagle as a symbol but also several Native American tribes and European countries did so in previous centuries. No one ever chose the harpy, however.

Diamondback Rattler

NEARBY MENACE

The eastern diamondback rattlesnake and the western diamondback rattlesnake, two separate species, are both U.S. citizens—and they kill more of their human fellow citizens than any other snake in the country. The bite of even a baby one can be fatal, turning tissues to jelly (to stretch a metaphor). Only about 15 percent of all snakes are poisonous, and these are a resounding pair of them.

The eastern diamondback slithers from North Carolina through Florida to eastern Louisiana, preferring piney woods, palmetto thickets, and old fields, but venturing into both saltwater and freshwater. The western diamondback swishes along from central Arkansas to southeastern California and everywhere south of that line, and it enjoys almost any terrain. Arizona probably has more of these species of snakes than anywhere else, and live rattlers, though not diamondbacks, have even been used there as a symbol of lightning in Hopi snake ceremonies. The rattlesnake is a powerful presence in America.

The best way to discover the presence of a rattlesnake is, of course, by listening for its rattle. Most snakes twitch their tails when aroused, but only the rattlesnake makes more noise than an ordinary tail would against dry leaves. Its rattle is constructed of hollow chunks of keratin locked together, with a segment added every time the snake sheds its skin. So, the louder the rattle, the bigger the snake. The rattle has evolved for the snake's safety and not its menace, since it enables the snake's predators to avoid it when they are not prepared and larger creatures such as buffalos and wolves to avoid stepping on it utterly by mistake. It is not wise to poke hand or foot under stones or into holes or bushes where this snake might be sleeping, either.

Rattlers can also be identified relatively easily by sight. The eastern diamondback is about 3 to 8 feet long, with diagonal white stripes on either side of its head. Its western cousin is a bit smaller, 3 to 7 feet long, and is notable for the black, gray, or white ring around its tail. In addition, both have the eponymous diamond shapes up and down their bodies, the western one's being less distinct.

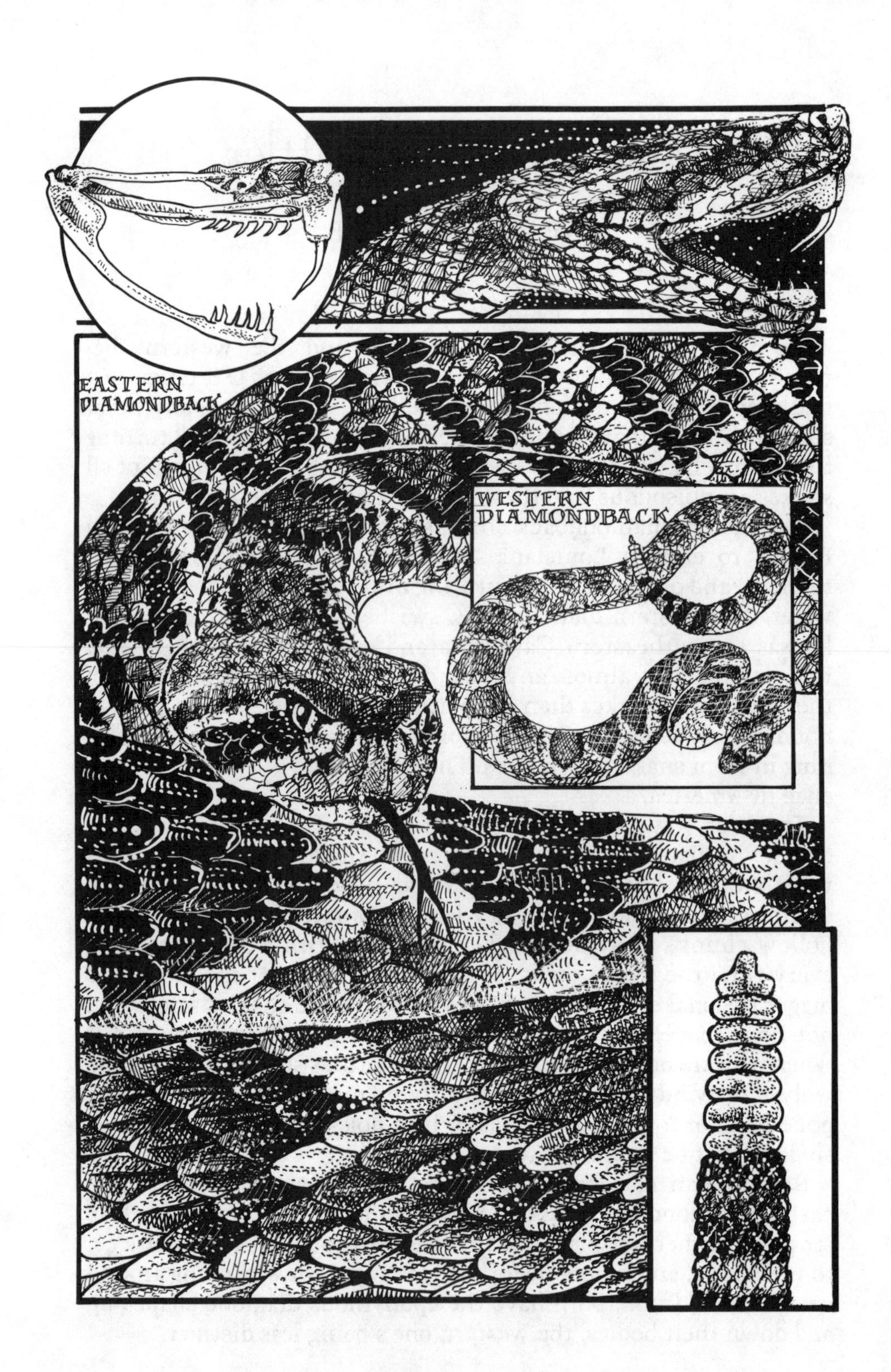
EASTERN
DIAMONDBACK
WESTERN
DIAMONDBACK

Rattlesnakes are known to be highly evolved creatures and may be as much as 12 million years old. Throughout all those years of slither, this snake has made other clever adaptations besides its rattle. Its fangs, for example, are highly specialized teeth with inner tubelets for moving venom. They are stored in the roof of its mouth until needed, whereupon the upper jaw rotates to bring them forward for biting. The snake has extra fangs on demand too; six more, in various stages of development, lie on each side of its face, ready to move down and become working fangs. The snake's poison can continue to flow even after its head has been chopped off.

Another unusual adaptation are its heat-seeking pits, a feature it shares with the others in its pit viper family. Two pits, one on each side of the face, enable it to detect a warm-blooded creature up to 1 1/2 feet away. A mouse, for example, emits about twice as much heat as the ambient air, and the snake can detect even a few thousandths of a degree in temperature difference if necessary. After finding and stabbing its prey, the rattler then uses another bodily feature—called its Jacobson's organ—along with its tongue to track down the victim as it makes its effort, usually a sadly short one, to get away.

The bite of a rattlesnake is noted for its painfulness, even among the pains made by other poisonous snakes. Swelling and puffiness in the wound are also especially marked, coming quickly after the bite. The venom is hematoxic, acting on the blood by breaking down blood cells so that they can no longer carry oxygen to the rest of the body. As the blood cells fail to coagulate, internal hemorrhaging proceeds. Cardiovascular shock and organ failure come next. The venom, which is just highly evolved saliva, is already preparing the body of the victim for digestion by the snake. The diamondbacks can kill a 200-pound person in an hour or two. One scientist who studies rattlers noted that just a glancing wound, from a single fang to his hand, made his hand and arm swell to the size of an inner tube. Even though he was rushed to the hospital by ambulance, it still took him four days of hospital treatment with antivenin to survive. Such treatment does work: out of about 8,000 people bitten by rattlers in the United States every year, 10 to 15 die.

The other good news is that rattlesnake venom, unlike that of the cobras, is not dangerous unless it actually enters the bloodstream; if it splashes on one's skin or is even swallowed, there will be no reaction—unless there is a cut, however small, on the skin, or a lesion, however tiny, within the digestive track. Since its power is through the victim's circulatory system, it is well to remember not to move around after an attack (which would stimulate circulation of the blood).

Diamondbacks are not the only rattlers in North America. There are about twenty-three other species, especially prevalent in the southwestern United States and northern Mexico. (Beware also, for example, of the timber rattler and the western pigmy rattlesnake; the latter prefers to live in blackberry bushes.) All rattlers have broad heads (for snakes), as well as the pit outlines on their faces. All eat lizards, chickens, frogs, insects, or small mammals. Everything is swallowed whole.

The diamondbacks are distinguished not only by their dangerousness but by their gracefulness. The males of many species enact a ritual combat dance. In it, one climbs up the other, and they then collapse gradually into a writhe and wrestle. Once fallen, they will begin the ritual again and will repeat it several times. It appears to be a territorial contretemps, but the rattlesnake dance is not well understood.

Jaguar

ONE LEAP ENOUGH

Its name means "the wild beast that can kill its prey in a single leap," and the jaguar is the most powerful predator in the jungles of Central and upper South America. Its range once extended from the southern United States to Argentina but now edges only as far north as Mexico. A jaguar needs an endless green world filled with wild game. At least there is now a reserve for these magnificent animals in Belize, where they often move along the area roads, trails, and waterways.

Jaguars are secretive, solitary animals, and are hard to see. They are mostly, but not entirely, nocturnal. Meeting only to mate, they otherwise keep apart in their partly overlapping territories by marking them with urine, feces, scrape marks—and growls. They are a rusty yellow on top, with black spots and bright green eyes.

Scientists who have studied jaguars remark on the hugeness of their paws and heads and the power of their jaws. At about 70 to 75 inches long, and usually 110 to 125 pounds, the animal is smaller than a lion or tiger and heavier than a leopard; they are, then, the third largest wildcat in the world and the largest in the Americas. Their heads are about 22 inches around, their jaws fierce with canines about 1 1/4 inches long—for crushing prey—and with carnassial teeth behind them, for cutting into it. The paw prints they leave can be 5 inches across. Their sound is a low, hoarse, coughing growl, or, more often, no sound at all.

Jaguars have been known to eat people, though this is rare. Their usual food is anything from an armadillo to a tapir, an anteater to a peccary. The latter, a wild pig, runs in herds and can sometimes turn the tables and tear the jaguar apart with their combined army of tusks. Jaguars occasionally eat capybaras (the world's largest rodent) and kinkajous (a jungle mammal) too. They usually stay somewhere near their kill to finish it (a word to the wise). When wild food is not available or jaguars are too weak to catch it, they will take cattle and then, very occasionally, progress to the owners thereof. Their hunting method is slink, hide, leap, then crush or puncture the prey's skull with their teeth.

A jaguar can be treed by a barking dog. And one will attack most readily when prey turns its back to run away. In the case of human prey, they often leap to within a few feet, then stop if you yell loudly at them. If you go to Belize for the jaguars, be prepared to yell your head off.

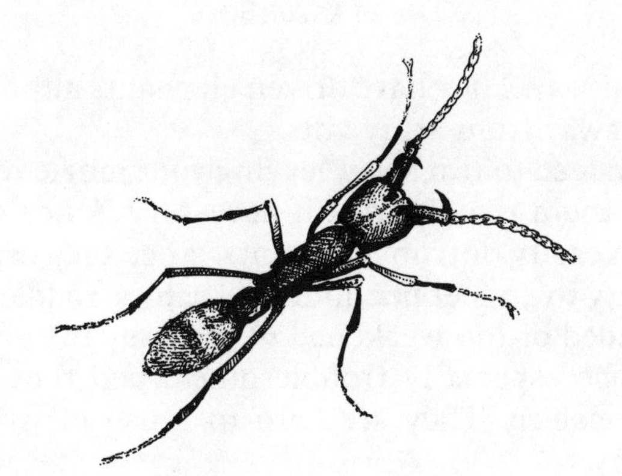

Army Ant

RETREAT AT ONCE

The Roman poet Martial wrote:

> While an ant was wandering in the shade of the poplar tree,
> A drop of amber enfolded the little creature.
> So that although in life it was regarded as nearly worthless,
> By its manner of burial it has now become precious.

Ants held a place of honor in the Roman mind, as we can tell from several other sources too. Aesop cast them as the wise ones in his fable about the profligate grasshopper who never stored grain for winter. And the Roman goddess of grain and agriculture, Ceres, was said to hold the ant sacred for the same reason.

But these people were not talking about the army ant of South America.

Army ants are quite ruthless. And, in their colonies of 20 to 30 million (an aggregation that weighs about 45 pounds), they are powerful. When they set out to hunt, it takes a lot to stop them, since they typically overwhelm prey with their numbers and sometimes crawl into eyes (to sting) and noses and throats (to suffocate). As a colony, they bite and sting hard and painfully enough to shock even a large creature into death occasionally. These ants relish eating other insects, but they have been known to kill lizards and rats, pythons too full to move away fast, crocodiles, even tied horses that are in their way. They have driven elephants almost crazy. Even anteaters run away from army ants.

The key is indeed to run away, leaving your jungle residence to be taken over by them temporarily, if necessary. A healthy, adult human being can easily outrun army ants, since they can only scurry at only about 25 to 50 feet per hour. But babies, toddlers, and adults who are wounded or too weakened to get away may be vulnerable. The ants can be especially troublesome when they surprise you when you are asleep. They are hard to brush off since there are always so many.

Here is your best army ant protection: once you see them coming, soak the ground ahead of the army with gas or paraffin and set it afire (the fire must be kept going for a while). If there is less time, use flame throwers. This will divert the ants toward someone else's jungle home or toward a nasty python perhaps nearby. But be very careful with the fire!

The armies must be an impressive sight. There are usually four parallel columns of the worker caste marching in groups of six to eight ants across, with guards at the edges. So many ants are marching that they can take several hours to pass one spot. Almost blind and with pale heads, they try to eat everything in their path that satisfies their carnivorous appetites. The marching orders are carried by pheromones, the chemical scents used by many social insects to communicate. It is the larvae they are carrying and the newly emerged adults who emit the pheromone required. When the signal fades, they cluster themselves into a ball, with passageways cleared between their bodies for their queen to move. She soon is laying her eggs, perhaps 25,000 in a few days. Soon enough the army is ready to march again, through the jungle—and through human nightmares.

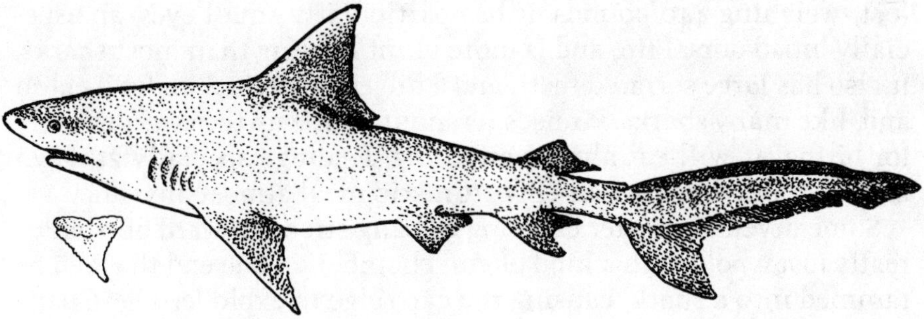

Bull Shark

ONE OF THE MOST DANGEROUS

Weasel shark, frilled catshark, basking shark, spatulanose catshark, collared carpet shark, crocodile shark, ornate wobbegong, blue-spotted bamboo shark, thresher shark, zebra shark, mallethead shark, hammerhead shark, megamouth shark, Iceland catshark, gulper shark, cookiecutter shark, rusty carpetshark, sand devil, daggernose shark . . . and there are many more. Some of these names, it seems, were divined or conjured in great fear. But of all the sharks that prowl the seas, the bull shark is one of the three most dangerous. (The other two are the Tiger Shark [see page 65] and the Great White Shark [see page 151].)

The bull probably accounts for more attacks than the great white and may be the most dangerous shark in the tropics. The reason is partly that it is found where no one expects it: in the Amazon River, Congo River, Bombay River, Brisbane River, and, yes, the lower Mississippi River, as well as in Lake Nicaragua. Of course, bull sharks swim the oceans too, all the waters off Indonesia, Southeast Asia, Africa, the east coast of South America and parts of its west coast, all of Central America, and up North America including California and South Carolina (even New England in the summertime). Attacks have indeed been confirmed in American coastal waters. These sharks hang out regularly around wrecks on the ocean bottom.

The bull shark is a dusky brown, gray, or black, and can grow to 11 feet, weighing 440 pounds. It has particularly small eyes, an especially broad dorsal fin, and is more blunt looking than most sharks. It also has large serrated teeth and a huge jaw tucked under its chin and, like many sharks, it raises its snout to line up the jaws perfectly for biting. It will eat almost anything, feeding especially at dawn and dusk but enjoying an opportunistic meal almost anytime.

Some divers feel safer carrying a "bang stick" to ward off sharks, really just a pole with a kind of gun chamber on one end that can be rammed into a shark, causing the cartridge to explode. These bang sticks must be kept perfectly waterproof in order to work, which is almost as hard as it sounds, and the diver must hit the shark at the right angle in order for it to go off.

Sharks are, believe it or not, in more danger from people, and from many methods beyond the bang stick, than they are a danger *to* people. Fishermen haul in about 100 million sharks a year, worldwide, some cutting the fins off, then dumping the shark back into the sea. There is no need for us to imitate the shark's own killing frenzy.

Piranha

BLOOD IN THE WATER

To readers who neither live nor swim in Amazonia, this fish is lots of fun. There are a myriad of bloody stories and horrifying, transmogrifying movie scenes about piranhas—relating in gory detail how a school of them quickly strips every tendril of flesh off a horse, a villain, and surely a nearby jungle explorer. Imagining this little fish's feats while standing near one in an aquarium store is quite amusing.

Not all of the tales about piranhas are true, however. One modern movie crew, trying to get footage, swam with a school of them—after pulling on wet suits—yet garnered only one small bite. Amazonian Indians swim with them, too. In fact, some tribes turn the tables and eat the piranhas. One tribe does, however, use them to clean the skeletons of their dead. But lots of piranhas are small and eat much more plant material over a lifetime than they do meat; in fact, only five of their twenty known species are carnivorous at all.

Some terror stories have been well documented, however. One capybara (a 100-pound rodent) was picked clean of its meat by a school of these fish, as have been a handful of excellent human swimmers and cattle. The largest piranhas, at about 2 feet long, are capable of nothing less. Blood gets these creatures going. In fact, if a school of piranhas is kept in the same water enclosure for one month, allowing their own various body waste products to build up, they will even attack each other. They are fast and they have very sharp teeth.

Smart South American swimmers keep a few things in mind: stay in clear water, don't thrash (which reminds piranhas of something's death throes), and be sure to avoid areas where there is blood in the water, for example, where slaughterhouse refuse is typically dumped into the river upstream. Most piranha attacks occur in muddy, rather shallow water, too, usually aimed at an already

wounded animal or a quite weakened swimmer. The attacks also cluster when the piranhas are especially hungry.

But piranhas have good points, even from a human perspective. some of these fish care for their young. And they also winnow out diseased fish among their prey species, thus averting fishy epidemics. They are indeed kings of the streams.

Snow Leopard

BEAUTIFUL, BUT . . .

These elegant, secretive animals live high in the Asian mountains in a land of snow cliffs and fallen boulders, above the tree line of junipers and wild peach. Snow leopards are almost the color of snow themselves yet have a leopard's black spots, making them practically impossible to see. They are also dangerously rare.

This leopard's favorite food is the bharal or blue mountain sheep, though they also hunt small yaks and musk deer, even pheasants, pikas, and marmots. They stalk like the cats they are, slinking along, then lunging, their long tails working as rudders when they leap. Typically, they make a large kill only once in every ten to fifteen days. After a meal, they groom their paws. Although snow leopards do not hunt people, these powerful animals will, if disturbed, tear your hand off in a second.

They are solitary animals, together only in the mating season and during the two years when they are being raised by their mother. During the mating time of January through March, they fill the deep valleys with yowls. The tiny babies are born in three months, to live a life of eight to ten years.

A typical day for a snow leopard involves a lot of moving around at dawn and dusk, with a lot of resting in between. They leap down cliffs and among boulders gracefully. They sleep in a different place every night, in the broken terrain of ridges and gullies where they are so well concealed. Since their territories are somewhat overlapping, they make sure to stay more than a mile apart from each other, keeping track by smelling each other's paw scrapes, urine scent marks, and even saliva markings on boulders. The same boulders are used as "highway signs" year after year.

The kingdom of the snow leopards lies in the mountains that stretch across Nepal, Bhutan, Tibet, Mongolia, China, the southwestern part of the former Soviet Union, Afghanistan, Pakistan, and India. No one knows how few are left, and their terrain is diminished more and more by human habitation. People naturally want their livestock safe from the leopards and want their villages to prosper. Though national parks exist, tourism has not yet truly come to the beautiful land of the snow leopard.

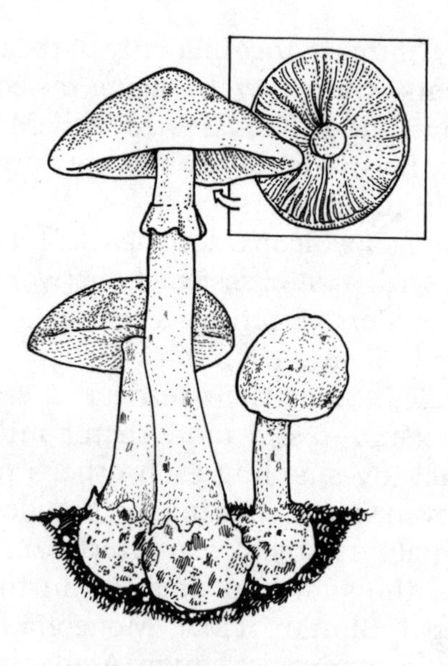

Death Cap

AN APPROPRIATE NAME

Also called death cup, this fungus belongs to the amanita family, which is responsible for more than 90 percent of all mushroom fatalities. Its symptoms are particularly tricky since they often don't appear for several hours, become severe, disappear, then kill.

Beginning usually ten to twelve hours (sometimes less) after a foolish feast, these mushrooms cause nausea, intermittent vomiting, abdominal pain, and plenty of watery diarrhea. This panoply of nasty symptoms lasts several hours, after which they all disappear and the victim feels fine for one to three days. Following that, however, come serious and sometimes fatal liver and kidney failure.

How to spot this unfriendly fungus: The cap, white or yellowish green to greenish brown, is 4 to 6 inches across and a bit convex then flattening; the stalk is up to 9 inches high. Under the cap the veil hangs down like a skirt, and on the underside of the cap lie white or pale gills. The vulva, like a sack, holds the base of the stalk. Its cap will separate easily from the stalk. Both veil and vulva may be mostly gone, but gourmets should still pick the whole stalk with the cap to check identification. Your life may depend on it. And do not rely on the fact that this mushroom smells like raw potatoes—it does not always.

Mushroom gatherers may see a little snail nibbling on a death cap mushroom and be lulled into thinking it a safe choice. Don't be fooled.

Fire Ant

EVOLVING TOWARD ENHANCED NASTINESS

Once upon a time, the fourteenth-century potentate Tamerlane had an empire that stretched from Russia to Arabia, Turkey to India. Once during his exploits, he had occasion to hide from enemies for several days in an old house. There, he became a bit discouraged and idly began to watch a single ant as it tried to climb a wall carrying something heavy. (You can't expect all the details from these old stories.) It fell back time and again—Tamerlane counted sixty-nine failures, it is said—but, on its seventieth try, it succeeded. The ant got up that wall. Inspired, Tamerlane resolved to go out and conquer Asia. Which, of course, he did.

Ants have had more roles in history. They inspired the very first human anesthetic, chloroform. This is merely a form of formic acid (a fluid made by ants) and was first made by crushing them into a quick squish of liquid. One more modern naturalist knew that ants were indeed attracted to this formic acid of theirs. So he painted his name on his own driveway in the substance. Soon he was able to go out and see that name come brilliantly, artistically to life, crawling in live ants.

Ants crawl the world at 8,000 to 10,000 species strong (and there are surely many more species remaining to be discovered). In fact, if all the creatures of the world were put on a giant scale and weighed, the ants, all rolled up together, would be 10 to 15 percent of that weight.

Even people who generally like ants hate fire ants. In the United States alone, every single year, some 67,000 to 85,000 people must see a doctor after being stung by these tiny red creatures. One or two Americans even die from allergic reactions to the ants' venom. The sting itches, then burns as the little pustules burst. And many sting at a time.

Moving beyond the parochial human level, fire ants devastate everything from orange trees to bird nests, fields where the grain seeds haven't even had a chance to sprout, even the insulation in traffic lights and airport runway lights (thus shorting them out). They can even kill cattle. And fire ants do eat other insects, too, tipping our scale of justice a tiny bit in their favor. They do all of this not only in the southeastern United States, where they established themselves after coming in from Brazil (probably in the hold of a ship); they have now spread to California and up to New Jersey. They are almost always the first ant to colonize an area.

Fire ant colonies, often 50 of them per acre of pasture with one queen per mound, were bad enough. But now these opportunistic crawlers have managed to evolve so that several queens, even up to 500, can coexist in a single colony. This is something heretofore unknown in antdom. They have evolved to living happily with a greater density of colonies per acre too, since these multiqueen colonies somehow lose their hostility to each other. Each of the queens lays eggs, of course, several hundred a day. As one U.S. Department of Agriculture entomologist put it, "We're seeing evolution in action. . . . It's wild, weird, wonderful stuff."

These new and evolving fire ants have taken over from the few unaggressive fire ant species that have always lived in the southeastern United States, and the newcomers are now said to occupy 400 million acres of land. All pesticide efforts have failed against their persistence, even though fire ants may now be the most avidly studied ant species of all.

They have indeed been extraordinarily hard to control. The ants prefer moist climates, and during one drought in the Southeast, they were flexible enough to get up and move inside people's houses. Since chemical pesticides have not only been ineffective but also have killed off other ant species more skillfully—thus allowing fire ants to colonize areas more thoroughly—efforts are now turning toward finding one of their predators or a disease from their old stomping grounds in Brazil. Entomologists are also researching biological controls and genetic alterations. This work will probably go on for a long time, and it is not likely to be entirely successful, even in the millennium.

The old verse from Proverbs says, "Go to the ant . . . consider her ways, and be wise." Also, beware.

Cane Toad

BIG APPETITE, BAD VENOM

This is a big, ugly, venomous carnivore and it breeds fast. Just ask the Australians in Brisbane and all around Queensland province, whose predecessors imported these South American toads in the 1930s to kill the beetles in their sugar cane fields. The toads are everywhere down there—at night they gather in circles at the streetlights for an insect feast, but they will also plunk right up and eat your dog's or cat's food out of its dish. They even eat carrion, baby birds, soup bones, plants, and cigarette butts off the sidewalks. When enough are run over on a road, the whole street stinks.

The toads breed all year long and plant black strings of eggs in any water that doesn't move. If the eggs from one pair of cane toads all reached adulthood safely, there would be, in one year alone, 60,000 more cane toads. There is usually one per square yard already in northeastern Australian backyards. Fortunately, they are cannibals too, at least sometimes.

These toads have two venom glands on the sides of their warty greenish yellow bodies that secrete fluid normally and discharge more when they are frightened. Out shoots venom, for up to 40 inches, and it can cause temporary blindness to people. When a snake or crow or dog or cat gets a cane toad in its mouth, that will probably mean death. A few people have dried out its skin, powdered it, and smoked it, since it can be hallucinogenic. Some have even licked the toads (alive) for a "high" that sounds equally low, as the above and could lead to a trip to the hospital. One Australian teenager even ate some of their eggs and died. The cane toad poison, called bufotenine, is taken from the family name of the toads.

So far, cane toads have not been known to kill very many people. In Australia, some people have even turned these constant companions of the yard and porch into pets. Without their venomizing their human friends, they can be patted and played with, provided the toads don't become frightened. People feed them to see how big they can get; typically, they range from 4 to 9 inches and can weigh up to 4 pounds.

Cane toads are, in the wild, nocturnal. The males have a loud call that is like a volley of deep flutes. They hide from the sun, which dehydrates them, though they can collapse their bodies without danger when deprived of water. They are scaly looking with hooded, primitive-looking eyes.

And don't be fooled: these animals go by the name "marine toad" or "giant toad" outside of Australia. Why not learn an ecological lesson here, too, about introducing species to a new place? Don't. And another lesson: live and learn, but don't lick . . .

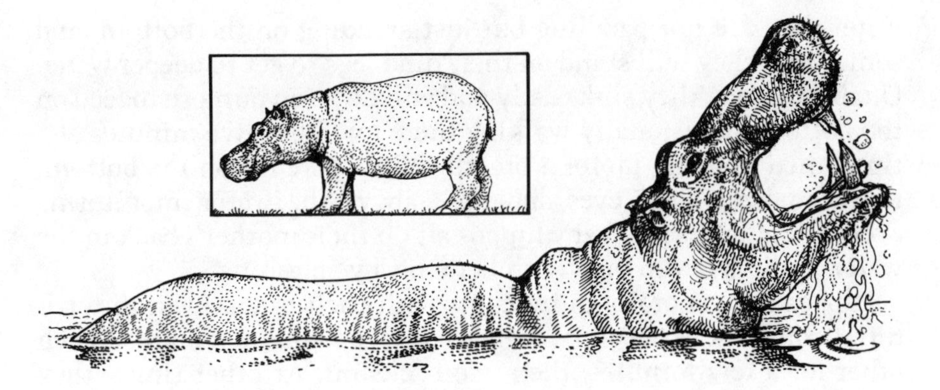

Hippopotamus

A CHUBBY KILLER

The British historian Thomas Macaulay once said, "I have seen the hippo both asleep and awake, and I assure you that . . . he is the ugliest of the works of God." A killer too. Hippos kill more people in Africa than all the lions, elephants, and water buffalo combined. They may look like big stuffed beer barrels, but they can outrun and trample people. Their canine teeth are dagger-sharp, though they rarely use them in attacks. Note that a hippo yawning could be evincing fatigue, but it is most likely delivering a stern warning. Sometimes they charge just as a warning, too, stopping within a few yards. (And sometimes not.) Of the two species of hippo, the pygmy is too small to worry much about.

Hippos eat 55 to 88 pounds of food a day to feed their 3,000 to 8,000-pound bodies, but that never includes human flesh. Vegetarians with sharp teeth, they go after grasses like a lawnmower, leaving the strips of land that are part of their territories pretty well mowed. This land feeding is an evening and night ritual, though they will come on shore in daytime to sunbathe if there is no one around to bother them. In the water, where they stay most of the day, hippos eat duckweed and a few other aquatic plants.

The hippo's name comes from the Greek word *hippopotamos*, meaning "river horse," but contrary to popular belief, hippos are not good swimmers. They are so big that when you see them in the water they are not paddling but just standing on the bottom, and sometimes they will stand on their hind legs to get to deeper water. Unable to float, they sink easily and will do so on purpose to feed on the bottom, occasionally walking there for about five minutes at a time, then hopping up for a breath. While perched on the bottom, they can keep their eyes and nose above the water, monitoring everything. In season, baby hippos sit on their mother's back in the water, and may even play with her a bit by splashing.

Hippos are generally quite calm with each other. Even in territorial conflicts between two males, they usually just stare at each other for a few minutes, then one backs off. At other times, they contest each other by swinging tails through the water while defecating and urinating, thus spreading their scent around—widely. Sometimes they end up bellowing at each other, mouths open, and throwing water. Very rarely do they actually fight.

A good territory to defend, in a hippo's eyes, includes plenty of shallow water (easier on the females when mating) and a narrow strip of land for their night grazing. They like to wallow in the mud to keep cool. (They have no hair but exude instead a reddish brown, sticky substance, which is a substitute for sweat and helps to cool them.) Hippo societies are quite stable, males maintaining the same territories sometimes for many years. The groups' numbers vary, with a few up to more than a hundred hippos in the same general area.

You have probably deduced that neither you nor your boat should make a territorial challenge to a male hippo. But neither should you disturb a female, especially during the first few days after her baby's birth. Like other ungulates (hoofed mammals), these babies will follow anything large during the first few days of life—and the mother does not want it to be you. She is also guarding her plump progeny against crocodiles, lions, and hyenas, all of which will go after hippo babies of all ages.

As long as it is healthy, the adult hippo has no enemies—except humans. People have long hunted them to get their tusks (easier to carve than elephant ivory), to keep them away from crops, and to eat them. They apparently have, under their 2-inch-thick hides, plenty of quite low-fat, delicious meat. Two thousand years ago, they used to waddle over almost all of Africa. Now their squeals, grunts, snorts, and evening "ooooooo" sounds are rarely heard.

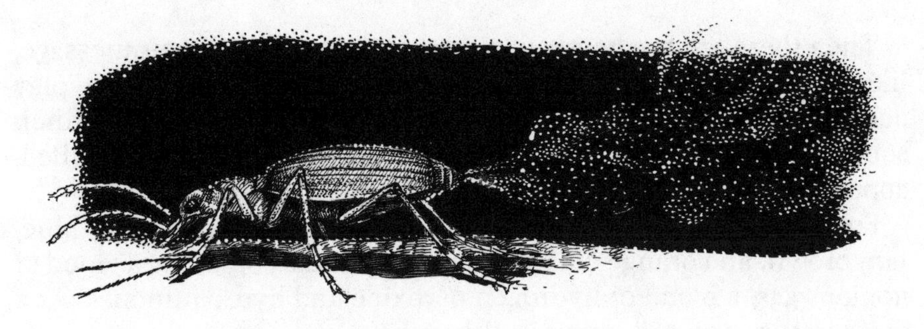

Bombardier Beetle

NASTY BLAST FROM THE REAR

Of all the dazzling array of animal species on this planet, fully one species in four is a beetle. So far, more than 325,000 different beetle species have been discovered, and there are probably many more. Beetles range from the huge African Goliath beetle to its littlest cousin, 8 million times lighter in weight—and everywhere in between.

Crawling and flying since the Permian period of 280 to 225 million years ago, beetles may have been the first insect ever to pollinate a flower. In fact, they still pollinate plants like the magnolia and water lily, which were among the world's earliest flowers. All were crawling and blooming, respectively, when the dinosaurs were rumbling around the earth.

Gone are the dinosaurs, but the beetles remain. And they have continued to grow in variety. Even their names are vivid: click, water penny, whirligig, ground, pleasing fungus, diving, longhorn, soldier, tiger, tortoise, stag, carpet, squash, and (my favorite) the confused flour beetle.

And many beetles have truly beautiful names, such as jewel beetle, ambrosia, beetle, goldbug, hercules, and soft-winged flower beetle. Others have wonderful reputations, such as the scarab beetle that inspired a million bracelets, the ladybug beetle (4,000 species strong) that eats up aphids and other pests, and one tropical species that gives off such a lovely light that women in Mexico and Central America have affixed them to their hair and clustered them in little bags on their skirts for the fanciest of parties.

Then there are the beetle names that convey a different message, like screech beetle, dung beetle, carrion (it even knows how to play dead to imitate its dinner), snail-feeding ground beetle (which squirts), and the grand master of the nasty insect squirt called, appropriately enough, bombardier beetle.

From the business end of its abdomen, this half-inch-long, blue, tan, brown, and orange creature defends itself by spraying a cloud of noxious gas, a blend of hydrogen peroxide and hydroquinones. At a temperatures of 212 degrees Fahrenheit, the larger "bombs" can cause real burns and make your skin turn brown. And these beetles are good shots.

Each blast of unpleasant effluent from this creature can be heard by the human ear as four to six volleys, but there are really 500 to 1000 fast pulses per second of emission. This firing method, which has been compared to German World War II buzz bomb technology, relies on two kinds of chemicals, stored separately inside the beetle, which are then combined in the spraying gland with enzymes, thereby to explode. The result is an impressive 26-mile-per-hour nasty blast—much faster than any anal muscles could achieve by squeezing.

Scientists studying this creature have tethered seven of them with tiny wires and pinched them gently with forceps. Courageously, they established that this beetle can turn its fanny nozzle in any direction, much like the turret of a tank. It keeps that gland sealed with tiny rear-end lips so that the explosion can't go backward, into the beetle. People often describe all beetles as "armored" on top and underneath. Certainly none is more like a war creature than the bombardier.

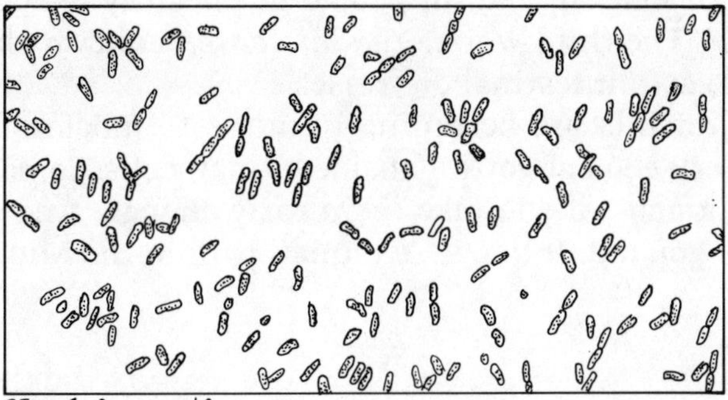

Yersinia pestis

Yersinia

SERIAL KILLER

While the Black Death and its related plagues have been blamed on the rats, and on the fleas on the rats, now the finger must be pointed at the bacteria in the fleas on the rats. Yes, Yersinia. This deadly bacterium, named for Alexandre Yersin, the French bacteriologist who isolated it in 1894 as the cause of the plague, grows fast and is spread easily.

Yersinia pestis is the most deadly of the three species in this family. Several times in history, it has murdered a significant percentage of the world's people, one-by-one, serial style. In fifteenth-century Europe alone, this bacterium killed 25 million people. (That makes lions and tigers and bears look certifiably harmless.)

Yersinia's method is to multiply within the esophagus of its host—the flea, say. The flea spits up plenty of the bacteria on its next blood meal—here the rat. The bacteria multiply more between the rat's cells too. Then the rat (or the flea) bites the person. The bacteria proceed quickly through the human body's lymphatic system. Within five to ten days, the person has septicemia, which may proceed to pneumonia. At this point, the plague is contagious through the infected person's saliva, and death usually comes to the first victim in fewer than an additional four days.

The second nastiest in the family, *Yersinia pseudotuberculosis*, is almost an anticlimax after the Black Death, but it does create diarrhea, emaciation, and, in people not healthy enough to begin with, death. The third worst, *Yersinia enterocolitica*, causes a full panoply of gastrointestinal distresses.

All the Yersinia are becoming immune to modern antibiotics. They are very good at getting inside human cells. Some medicines do work, though—if you take them early enough. And thank your lucky stars you didn't live in Yersinia's path in the Middle Ages.

Water Moccasin

A.K.A. COTTONMOUTH

Out of the more than 2,500 species of snakes that creep through the world, about 650 species are venomous, and not all of those fatally so. The water moccasin, which lives in the United States, is one. It can also be particularly belligerent.

Part of the pit viper family, this snake is a denizen of shallow lakes, slow streams, and swamps from Virginia to Florida to Mississippi, up to southern Illinois, back down to parts of Missouri and Kansas, and then west to Texas. Look for one basking on a log or rock near the water in daytime, and watch for it even at night if the weather is warm.

Water moccasins are from 3 to 6 feet long, with broad, flat heads more distinct from their necks than most snakes' heads are. The head is usually a dull brownish black with either a pronounced or a faint white line that extends back from its eye.

This snake can be distinguished from harmless water snakes because it will not flee from an encounter. The water moccasin looks impressive when it takes a vertical position and opens its fang-decorated mouth. As the eponymous cottonmouth, its other name, it has a big white mouth lining. If you are looking at that, beware.

These snakes can be lethal, and even when not so, their bites cause severe tissue damage and scarring, with a recovery that can take several weeks or months. Along these lines, ponder the scientist who has been bitten by them (as well as by many other snakes) several times. This man has developed an immunity to many venoms—the hard way—and deliberately. He is sometimes called to give his own blood as medicine to victims when no other antivenin is available. This scientist, like other herpetologists, actually milks snakes for their venom; he supplies viper venom, for example, to pharmaceutical companies developing not only antivenin but anticlotting and cancer medicines based on snake venom.

The dance of the male water moccasin, a ritualized combat almost all of which occurs in the water, is particularly intriguing. One naturalist standing in his yard in Florida scrutinized a nearby swamp only to see two of the males gliding along, twisted together, heads side by side, then crossing and recrossing their necks without touching each other. The smaller snake then circled the larger one, touched it with tongue and snout, next the two looped together and rose higher out of the water. After several repetitions taking nearly an hour, the larger snake began a crazy zigzag, the smaller one chased, and the larger one hurried to hide itself on the shore. No one truly understands this dance . . . and probably no one wants to get any closer to it either.

To end on a lighter note, here is a Shakespearean incantation of a snakely sort:

> You spotted snakes with double tongue,
> Thorny hedge-hogs, be not seen;
> Newts, and blind-worms, do no wrong,
> Come not near our fairy queen.

If the snake is a cottonmouth, not near the rest of us either.

Killer Bee

A CLICHÉ COMES TRUE

"**H**ow doth the little busy bee/Improve each shining hour,/And gather honey all the day/From every opening flower!" No one sings lyrics like these about the killer bee!

Killer bees may be the only insects to have become cliches, and, like many cliches, part of the story is woefully true. They do swarm over people—when have they been even slightly disturbed, even by a person walking past their hive—stinging them about 24 times per second. This was actually measured by scientists with a gizmo called a *tempertester* or *sting-o-meter*, and it compares to 4 stings per second from "regular" bees. Yes, they chase you and will fly in pursuit for at least half a mile, much more if you happen to have taken their queen with you. They have, in fact, killed 700 to 1,000 people in the Americas in the last thirty to fifty years—with anywhere from 1 to 500 stings at once.

The bees are now advancing within the United States, having proceeded north from Brazil at the rate of 200 to 300 miles per year; they left Brazil in 1956 (thanks to the accidental release during a bee experiment of 26 queens and possibly 250 drones) and, about 150 bee generations later, are pestering Texas and California. Their first stings have already occurred there.

Killer bees are more formally called Africanized bees because they are a hybrid of Brazilian and Tanzanian bees (the brilliant idea of the scientist whose technicians allowed them to escape). Their aggressiveness evolved in Africa where plenty of creatures, including the honey badger, have always been after them. And they are even worse for America's crops than they are for people. Many crops depend on beekeepers and their nice bees, brought in to pollinate the plants. The killers wreck this sweet arrangement. They get into the hives of the nice bees to mate with them but don't let the good guys into their hives; this aggressive mingling, which was once predicted to mute the killer bees' fierceness, has not worked too well. The killer bees tend to overwhelm the gene pool. As one scientist says of a given area, "Complete Africanization results in two to three generations." Only some new efforts at producing large numbers of "nice" drones to mate with killer queens are showing some promise.

The new killer bees do not like to find all their pollen in one area the way the old, good bees do either. In fact, in Africa they are considered nomadic species, so crops don't get pollinated. The killer bees produce less honey too, hurting that small but tasty industry. And, of course, they won't let you get what they do make. If this scenario seems overblown, note that crops and honey production in northern South America and in Central America have already been damaged. (In Venezuela, for example, the honey crop was cut by more than 75 percent.) Of course, part of the reason may be that terrified beekeepers simply abandoned their hives when they found out what was in them.

Killer bees are hard to distinguish from honey bees at first sight. They are just 10 percent smaller and a tiny bit darker. Each one actually holds less venom (but then they all sting at once). They have shorter life spans but swarm more frequently to find a new hive. The only way to tell if that bee coming at you is a killer bee or nice bee is to use a microscope—or see how much quicker they are to anger (30 times faster), or longer to forget that anger (30 minutes vs. 4 minutes). It is not clear how far north they can survive the winters. We'll know soon enough. Meanwhile, I wouldn't watch any of those grade B movies—you'll be plagued with nightmares.

Grizzly Bear

DEADLY AND MAGNIFICENT

A northwoods Indian proverb says, "A pine needle fell in the forest. The eagle saw it fall. The deer heard it. The bear smelled it." With a sense of smell better than a bloodhound's, lumbering legs that can clip along at 36 miles per hour, paws that make footprints 17 feet apart when it runs, and a muscular bulk of 1,000 pounds or more, the grizzly is considered among the world's most dangerous animals.

Grizzly bears are omnivores, and a small swatch of their menu would constitute a real cafeteria: huckleberries, blueberries, biscuitroot, kinnikinnick, glacier lily bulbs, sedges, elk thistle, violets, yarrow, muskrats, voles, carrion, garbage, dog food, dogs, dead whales on the beach, salmon, trout, weakened caribou, moose or elk or buffalo, and—sometimes—people. About 80 percent of the grizzly diet is vegetable and fruit, but that leaves 20 percent. As John Muir put it, to a bear, "almost everything is food except granite."

In the days of their glory, when perhaps 100,000 of them roamed North America, grizzlies chased both a surprised Lewis and later a surprised Clark in the west . . . and almost got both of them. Tragically, the numbers of these exquisitely dangerous bears have fallen with the ungreening of America. Descendants of a bear-like creature that roamed Europe about 25 million years ago, and of the first true bear there 3–8 million years ago, the bears moved to the New World and grew huge to adapt to the tundra environment as the glaciers retreated. They ruled North America for thousands of years, dominating the land from the Bering Strait south to northern Mexico and east to the edge of the prairie. There are now only 10,000 to 15,000 of these bears in all of Alaska, 18,000 to 20,000 throughout western Canada, 400 to 700 in the wilderness in and near Glacier National Park, and fewer than 200 in the Yellowstone area.

Surprisingly few people have actually been killed by grizzlies. During the twentieth century, only about fourteen people so far have been mauled to death in the United States (not counting Alaska). In Glacier Park, just one visitor in 1 1/4 million or so actually sees a grizzly (of course, that counts all the people who never

get out of their cars). In the Glacier area, a grizzly has a range of about 190 square miles—so if a bear is sighted at the far end of the park yesterday, it can be sniffing at the food here in your picnic basket today. Their territories vary from 10 to 1,000 square miles depending on the quality of the food and shelter in a given area.

The fearsome grizzly is distinguished from the mostly peaceable black bear not by the color but by the hump. Only the grizzly has a huge shoulder hump, and it is all muscle; some of it is connected to the powerful jaws.

When a grizzly is a cub, it is tiny and almost helpless. Born in January in its mother's winter den and weighing only about 1 pound, it is both blind and bald. Out of the den in April, the now-cute young bears have been seen sliding down snowy hills on their fannies, then running up to sled again on their stomachs, and even throwing snowballs. They remain under the careful tutelage of their mothers, play-fighting and play-hunting, for two to three years out of their natural lifetime of thirty years. Never interfere with this domestic scene by getting between cub and mother, or between bear and buried food (a carcass). Bears make nests near these caches to stay near them for a few days. Grizzlies take a nap around 2:00 P.M.; if the temperature is higher than 72 degrees Fahrenheit, they usually rest all day.

There are many tips for avoiding a fatal confrontation with a grizzly. First, and best, is to ring "bear bells" (just jingle bells) incessantly or sing lustily as you hike in grizzly country. Do this even if your children consider it embarrassing. The bells, jangled even from quite far away, alert the bear to stay away. Avoid wearing perfume, avoid hanging around in berry patches, avoid sleeping in the same clothes you cooked dinner in, avoid camping out entirely during your menstrual period or when you have a major wound, and avoid sex or urination in or near the tent.

If you do come face-to-face with a bear, never run away; bears chase anything that flees, a hunting strategy that has served them well. (They also enjoy eating 20 pounds of the meat at once, if they can.) Climb a tree 30 feet high if you can. This is higher than the biggest bear can reach a paw when on its hind legs, and grizzlies don't like to climb trees. Or open a big umbrella suddenly and loudly. (This works, say people who have done it.) Never stare at the bear, since that means aggressiveness in the bear world. Also, make no sudden movements, just back up slowly and speak calmly and respectfully to the bear. Some people have successfully played dead or even shouted and swore until the bear left. Still others have done well by carrying a squirt gun full of water and red pepper and aimed for the bear's eyes, to create a temporary itch. And, last, if the bear does indeed charge, sometimes it is only bluffing.

These bears are territorial and, if you are in the territory of this juggernaut, congratulations. Bear country is wonderfully wild, worth it, and there should be more of it. Hike respectfully. The grizzly could be the most magnificent threatened species of all.

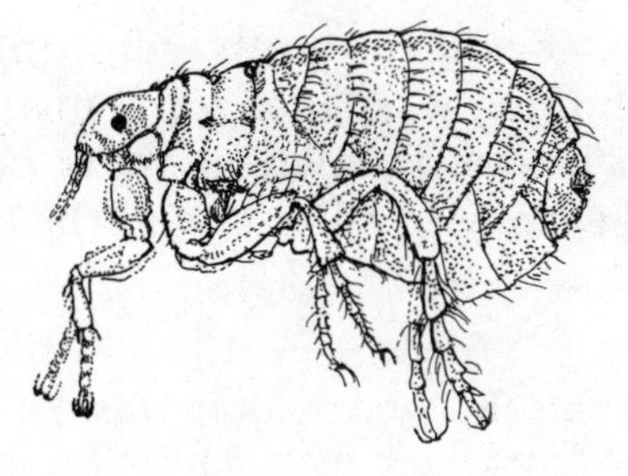

Flea

HIGH JUMPER

See the little flea decked out in its tiny gold collar, pulling the miniature ivory coach along the table top, complete with a carved king and queen along for the ride. Watch the companion flea in its collar of silver, leaping high, lifting a small but perfect royal crown behind it. If you lived in sixteenth-to nineteenth-century Europe, you might have been able to attend such a regal (and legal) flea circus. Real fleas were decked out for performances, and people paid real money to view their antics. The owners fed their employees by letting them suck a little blood from their arms.

These flea trainers took advantage of what fleas can do anyway, which is drag more than twenty-five to fifty times their own weight and jump terribly high. Their brute strength, actually common among insects, comes partly courtesy of the better leverage that muscles have when attached to the outside of the skeleton. (Humans have muscles attached to the inside of their skeletons.) The jump, for which the flea would win the Insect Olympics, can take the flea as high as 13 inches, this from a creature just one-tenth of an inch "tall"; in this trick, it is assisted by its relatively large muscular legs equipped with resilin, a kind of natural rubber whose elastic properties have not been equaled yet in any artificial substance. Outside of the circus ring, one researcher conducted a study in which he induced one such tiny athlete to keep jumping, once per second, for three whole days. For entrepreneurial (and exploitive) careers in the entertainment industry, one need look no farther than this.

Attitudes toward fleas in pre-twentieth-century life seem to have been more relaxed, if not uniformly admiring. This is probably because living with them was taken for granted. The seventeenth-century poet John Donne (known, it must be admitted, for his offbeat metaphors) even used a flea bite as the center of a romantic poem:

> Mark but this flea, and mark in this,
> How little that which thou deny'st me is;
> It sucked me first, and now sucks thee,
> And in this flea, our two bloods, mingled be;
> Thou know'st that this cannot be said
> A sin, nor shame nor loss of maidenhead,
> Yet this enjoys before it woo,
> And pampered swells with one blood made of two,
> And this, alas, is more than we would do.

Of course, people in those days sometimes wore sticky collars around their necks to trap and kill fleas too, and also scattered the plant "fleabane" around the place for its equally alleged murderous properties toward the little creatures. It is worth noting, too, that the fact that the words *flea* and *flee* sound alike is not a coincidence: the insect's name derives from the Old English verb meaning "to flee." For good reason.

We could go on with flea fun forever, but the time has come for more cautionary tales. These fleas have carried bubonic plague, typhus, and tapeworm, and, as only one example, they caused the Great Plague of the years 543–544, which spread from Ethiopia to Byzantium to Europe; during this time, fleas were responsible for 10,000 deaths every day in Byzantium alone. At the time of the Black Death, they cooperated in the killing of 13 million Chinese and some 75 million Europeans. Rat fleas are usually the plague carriers—they spit up the relevant bacteria as they bite the next rat, or person.

There are some 1,500 species of fleas and each has its preferred host. In Antarctica, there is even a penguin flea. The brushy combs on their little bodies have spaced "teeth" that match the width of the fur or hair they have evolved to grab; the combs are the largest, for example, on the porcupine flea. (The size of the flea's body is less matched to its host—the shrew flea, for example, is a full one-tenth the size of the shrew itself.) Baby fleas are quite efficient at finding their host once they have entered adulthood; one scientist marked

270 rabbit fleas, let them loose in a 2,000-square-yard meadow, trapped all the rabbits a few days later, and found a full half of "his" fleas on them. (This man should start his own circus.)

Fleas are very flexible about their diets. First of all, they can live for many months without a meal if they have to, even when newly emerged as adults. And they will change hosts to get their food, although it may not be their favorite blood menu. People who notice three little flea bites in a row often wonder why; the bites were probably caused by a cat or dog flea who didn't like your blood much on the first bite, tried again, but finally gave up. The fleas who usually end up biting people include those from our household pets, and also the rat and chicken flea.

There is also such a thing as a human flea, though it will also parasitize badgers, skunks, squirrels, dogs, and pigs. I guess we should be known by the company we keep.

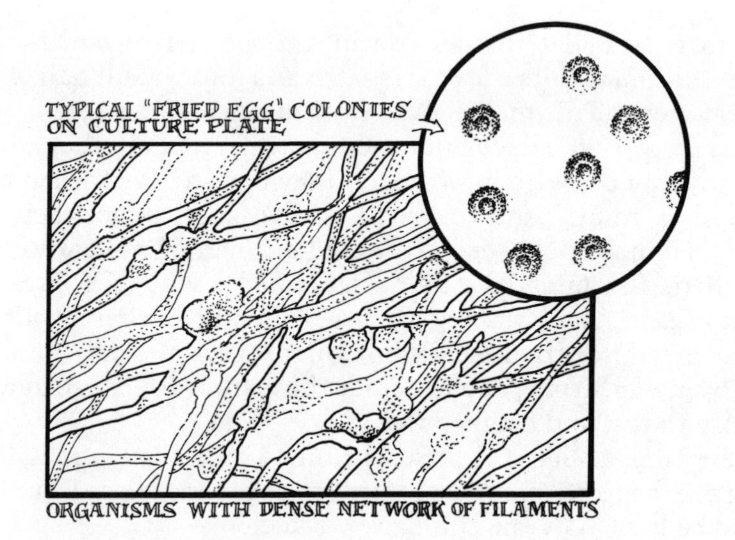

Mycoplasma

BAD BACTERIUM

There are sixty-nine different species of this microbe, the smallest and simplest creature that can live without a host. Now classified as bacteria, they inhabit the world in their remarkable, horrid, and invisible diversity. One species of mycoplasma causes arthritis in rats, another infects ducks, several hit on cattle, and one each on squirrels, pigeons, horses, pigs, guinea pigs, and cats. Among the several that infect dogs, one can move from the dog's throat to its human family. (So avoid kissing your dog.) Two live in people parasitically but seem to do no damage, four more very occasionally create premature labor or kidney disease, and one creates a nonfatal pneumonia. They all grow inside their hosts in colonies shaped like fried eggs (once arranged on a laboratory slide).

Then there is *Mycoplasma hominis*, one of the worst of this bad batch from the human perspective. This species clings to the lining of both the male and female urogenital tract to grow in the same fashion as its fellows. It is what inflames uterine tubes, forms pelvic abscesses, and creates septic arthritis. As its toxic by-products build up, it can do extensive tissue damage and lead the immune system into making an inappropriate response. Engineer, thusly, of the quite woefully common pelvic inflammatory disease, it can also kill fetuses in the uterus and kill the mother in postpartum fever.

Yet another dreadful mycoplasma species, *Mycoplasma fermentans* (also called the incognitus strain) has recently been discovered. This one can kill nonhuman primates, and can induce lesions in human lymph nodes, the spleen, liver, adrenals, heart, and brain, possibly resulting in death. This species also somehow prevents the body from stopping the spread of the HIV virus, thus creating full-blown AIDs. Its tracks have now been found in many patients.

Krait

ALMOST PAINLESS

The famous Swedish scientist Carl Linnaeus wrote in 1797 that, "Reptiles are abhorrent because of their cold body, pale color, cartilaginous skeleton, filthy skin, fierce aspect, calculating eye, offensive smell, harsh voice, squalid habitation, and terrible venom; wherefore their Creator has not exerted his powers to make many of them." Even coming from a man known for his studies of flowers, and in a relatively snake-free country, this seems unnecessarily—and wrongly—harsh. (Besides, there *are* plenty of reptiles in the world.)

When it comes to "terrible venom," though, he was right about the kraits. These Asian snakes, about twelve species strong, are closely related to the cobras. One of them—the many-banded krait (also called the banded krait)—has venom almost twenty-three times as deadly as even the king cobra's. It is one of the most toxic snakes of the whole elapid family.

Kraits tend to live near people's houses or their rice paddies. But most are nocturnal and seem to stay in a sort of stupor during the daytime, usually in their burrows, so they tend to be hard to provoke. Depending on the species, their foods of preference are lizards, mice and rats, fish, and other snakes (and not people).

Most kraits are 3 to 5 feet long, with two of the species up to 7 feet in length. They have small, flattened heads with small, dark eyes, and are glossy, cross-banded, and, of course, fanged. The banded krait is found in greatest concentrations in the East Indies; and the other most dangerous species, the common krait, is common in India, throughout Southeast Asia, and China. (Keep in mind that Southeast Asia has the richest variety of snakes anywhere in the world.)

If they bite, kraits are sometimes able to hold on hard with their jaws, chewing on the flesh and thereby administering plenty of their neurotoxic venom through their fangs. The krait bite creates almost no pain (unless the snake reaches that chewing stage) and very few symptoms—for the first three to five hours. Then, however, the victim can experience intense abdominal pain, shock, coma, and death, quite quickly. If not treated promptly, the fatality rate per bite can be quite high.

The seriousness of any snake bite depends on many factors, and they are worth considering as one ponders the krait: the bite's depth and location (it is less dangerous to be bitten in the extremities); the length of time the fangs are sunk into the flesh (the longer, the more venom flows); the size of the snake within a species; the condition of its fangs and venom glands (if it has recently bitten something else, the gland will be at least partly empty); what else is in the snake's mouth (bacteria from its last meal, for example); allergic sensitivity, age, and size (though obesity is hardly a complete defense); and how much help the victim gets and how fast. Iceland and Hawaii are great vacation spots—they have no snakes.

Elephant

A BIG HAPPY FAMILY

A baby elephant weighs about 250 pounds at birth, but for several months it is awkward enough to trip over its own trunk and even get it stuck in the riverbed while snorting up a drink. The baby sucks its trunk, too, like a child's thumb. As adults, elephants can use these trunks (with their thousands of muscles) for everything: digging for water, pulling the big baby up a slippery riverbank, snarfing up 300 to 350 pounds of food a day, and trunk-hugging with other elephants. They also curl up the trunk between their tusks to rest.

Elephants can kill, not only by charging or trampling but by hitting with one swipe of the trunk. The charge is sometimes a bluff, but not always, accompanied by much flopping of ears.

There are only two subspecies of elephant left now, and neither can be considered much of a hazard: the Asian elephant numbers fewer than 20,000; the African only about 600,000 out of what was once millions and millions of creatures. This animal's ancestors once used land bridges to lumber onto every continent except Antarctica and Australia.

Elephant numbers are tragically few not so much because of destruction of their habitat (although that is happening too)—African elephants occupy swatches of territory within about thirty-five different countries—but because of the appeal of their exquisite tusks. The demand for that ivory bracelet or figurine is murderous. Kenyan officials have managed to get a ban on ivory accepted throughout most of the world, drying up the market, and prices for this living material have fallen drastically. They have fallen so low, however, that even minor African chieftains can now afford ivory and have begun to create a new, though smaller, market. At least the Japanese have invented an ivory substitute made out of whole eggs and milk. However, it is not being used extensively.

Those who doubt the elephant's value might want to ponder the sophistication of its life. Elephants have an elegant, elaborate social structure based on groups of female elephants, most of whom are known to be related, along with their offspring. (All the females watch over all the young.) These groups associate with other such groups and, when they meet, will exchange greetings by curling trunks for fifteen or twenty minutes. This larger "bond group" can number as many as 400 elephants. Even within the larger group, they will come to each other's aid when signaled.

The males roam alone, leaving their small female groups at about age thirteen, then visit various such groups again to mate when "in musth," which begins for them at about age twenty. This sexual state makes for dramatic and deafening trumpeting, ear flapping, urine dribbling, and also invades a special eye fluid dribbling down their faces. Fights between males are common during this period. The females create great commotion too.

When temporarily apart on their range, or when searching for mates, elephants call to each other over several miles, in sounds mostly too low (down to 14 hertz) for us to hear. They have, of course, plenty of audible roars, trumpets, snorts, screams, bellows, and rumbles. In fact, elephants have about thirty known calls.

Intriguing rituals connected with death have been observed occasionally among these very subtle, intelligent creatures. When an elephant in the group dies, others have been known to bury it by gently placing the branches of trees and dirt over its body. And when they encounter the bones of an elephant later, some have been seen to carefully mouth and pass the bones around, then place them back on the ground. They do not do this to the bones of any other species, and their behavior occurs in which scientists describe as a state of "quiet tension." Also some elephants were reported to have taken a wounded baby several miles directly to a forest ranger's office, seemingly for protection from a tiger that had attacked it.

They can be very playful too, and even adult elephants pirouette around and toss logs in the wild. But it is a good idea to keep away from males in musth, from females who are mating with them, and from their offspring.

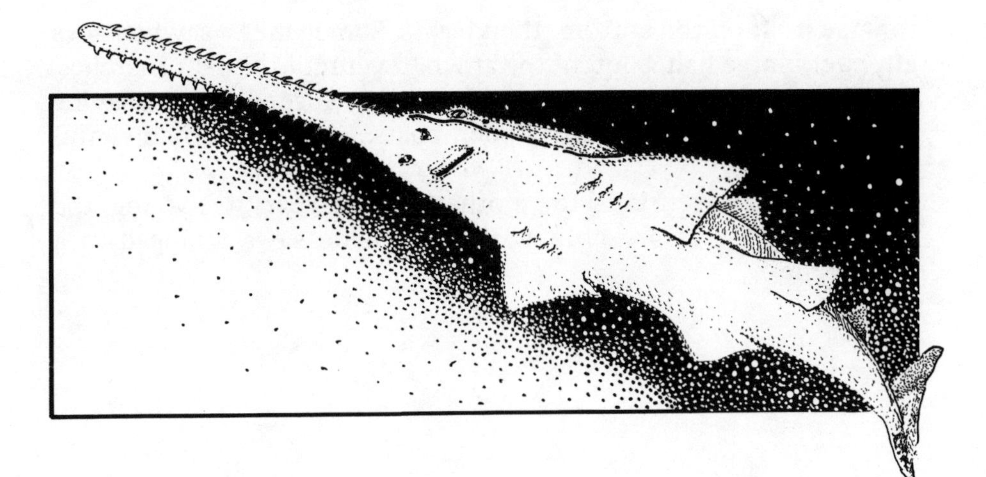

Sawfish

A NATURAL TOOLKIT

Pliny the Elder, everybody's favorite source of fanciful monster information, wrote that sawfish in the Indian Ocean could reach a length of 300 feet. Fortunately this is not true. But even at their maximum length of 20 or 30 feet, and weight of up to 1,000 pounds, they are distinctly dangerous. As members of the family that includes both rays and sharks, they have been known to attack swimmers.

Sawfish are very common. They havoc through all tropical seas, lakes such as Lake Nicaragua, and up rivers that drain into all of the above. There are six species, all with flattened heads, teeth inside their mouths for double trouble, and notched saws that extend from their faces. They swim like sharks but, like rays, have their gill openings on their undersides.

The problem is, of course, that eponymous saw. They use it to dig into the ocean, lake, and river floor, and for defense, but its main use is as a slasher. The sawfish often swims into a school of fish swinging its appendage, eating the fish pieces that result. (The bits that fall aside go down to feed the anemones and other bottom creatures.) If a big chunk of fish gets stuck on its saw, the fish goes down to wipe it off on the bottom, then feasts. Sometimes a sawfish goes after very large fish, content to eat whatever mouthfuls it snatches, or the entrails that tumble out of the wound. It goes without saying that fishing nets and small wooden boats are no safer than swimmers.

In case you are curious, the mother sawfish bears live young; the 2-inch-long babies are born with their little saws wrapped in a membrane.

Toadfish

JUST PLAIN UGLY

These unattractive bottom dwellers lie well camouflaged in their sand, mud, and seaweed homes, decked out in green, brown, yellow, or a muddy medley of those colors. They have big heads, big mouths, and are about a foot long. Toadfish are not actually as deadly as the stonefish they resemble (see page 44), but they are a lot more common.

Toadfish have four main ways to irritate people. First, they can bite off a finger; usually sluggish, they can become fiercely territorial when disturbed and are quite strong. Second, they have spines on the sides of their heads connected to a venom gland that produces painful stings; these stings are not life-threatening, but victims must watch out for secondary infections. Next, their flesh is poison . . . should anyone be so misguided as to want to eat one of these fish. And last, they can grunt loudly enough to scare you witless if you handle one or swim near it, and loudly enough to keep you awake at night if you live near their habitat. At 2 feet away, their boop-grunt is as loud as a subway train.

Toadfish live along the East Coast of the United States and down to the West Indies, in the Gulf of Mexico, up the American West

Coast, in the Mediterranean and the Red Sea, and throughout Southeast Asia. Most species are ocean dwellers, but others have chosen estuaries and even freshwater habitats. Though they generally prefer deep water, they come into shallow water to spawn every summer. There, the male uses his swim bladder, shaped like a heart (no less), to make the loud noises that attract the female toadfish. (Fish do not have vocal chords.) Once mated, he is the one to guard the eggs for several weeks. When the young hatch, however, he will eat the ones that don't swim away fast enough.

This charming creature has yet another oddity: a row of lit-up spots on its side. They are used, scientists think, to guide the female to the male's nest under the rock. This bioluminescence is found in the toadfish species who feed on small crustaceans who themselves produce light (as do other deep-sea creatures who are unrelated to the toadfish). The toadfish is a light and sound show all in one.

Horned Toad

NOT YOUR BEST PET

Though the Asian horned toad is sometimes kept as a pet in Asia, it is a thoroughly ugly creature that can make an unpleasantly metallic "ching" sound. It is hard to know why anyone would want a Texas horned toad around either. And they don't do well in captivity.

The Texas horned toad is part of a family of fifteen species found only in North America. They all live in dry areas, even sometimes mountains, and come out of their burrows only in warm weather. The Texas species, actually a lizard and not a toad at all, lives in Texas, New Mexico, Arizona, Colorado, Kansas, Oklahoma, Missouri, Arkansas, and Louisiana. At 4 to 5 inches long, it is a gray, brownish tan, or dusky reddish or yellow, always with little horns down its back and bigger ones at the rear top of its head.

Do not upset this lizard creature. When it feels threatened, it puffs itself up. Then it stands on its tiptoes, opens its mouth, and hisses. Its blood pressure has been rising all this time (as has, probably, yours) and by this point the capillaries near the corners of its small, black eyes may be swelling and getting ready to rupture. If it is indeed this upset, it directs blood from burst blood vessels right at you—accurately up to about 7 feet. This defense can squirt out up to one-fifth of its blood supply (which the toad quickly replenishes).

Great White Shark

"JAWS"

The most dangerous shark in temperate waters, the great white is the only shark to have made the cover of *Time* magazine, on June 23, 1975. This is "Jaws." Most people think this shark is white, but watch out: its top half can be a deep blue or gray-green, gray, even brown, with the white on the bottom. It is especially muscular. And it is the only shark (indeed the only fish) that regularly lifts its head out of the water to look for prey.

The great white shark is looking for seals and sea lions, which it enjoys, along with salmon, tuna, large turtles, and dolphins. This is a large creature—up to 25 feet long and weighing 3,500 pounds—and it eats large things. It captures its prey using excellent vision, hearing, and smell, as well as sensitivity to both electrical and magnetic fields. Sharks may well be the most electrically sensitive fish in the sea.

To a great white, a person is not very differently sized from its ordinary menu items, and we can be easily confused with seals when we don our black wet suits; with long-flippered sea lions when we climb on our surfboards; and with turtles when we are on boogie boards (similar to small surf boards). Attacks on people are usually mistakes or warnings to get out of shark territories, but humans do seem to taste good to them.

A typical great white shark attack involves one large bite first. The shark even rolls its black eyes back in its head to protect them from being poked in the melee. Then it just backs away to wait for the prey to die from loss of blood, shock, or damage to vital organs (or all of the above). This fish is known for its patience and will circle for quite a while. Absolutely nothing scares it away. Then it simply finishes the meal.

Sometimes great whites swim up to people—and then just turn away. But, off the coast of southern California alone, they attack three or four people a year. The shark's numbers are increasing in this area, and 1991 was a very bad year for surfers and fishers along the Carmel/San Francisco coastline. Most people survive the attacks but with body parts missing. In fact, one California man was diving for some abalone for dinner when a great white attacked. His head was actually inside the shark's mouth, but he was able to pound at it with his spear gun until the shark spit him out. Attacks have been documented off New England, too, up the west coast to Washington state, and regularly around Australia, New Zealand, and South Africa. This shark's range can even extend along all of America's coasts (occasionally up to the southern edge of Alaska), off parts of South America, in the Mediterranean, then south to areas of Africa and north to France, around Korea and Japan, and up to northern Russia. A man diving for shells was bitten right out of his wet suit in the Japan Sea recently, for example, and did not survive. They are not, however, common in the tropics. These sharks prefer deep water (below 100 feet) but do not always go with their preference.

Sharks, in general, have been patrolling all the oceans in much their present form for more than 300 million years. They are a success story of evolution, even having evolved protection against cancers. Most are peaceful plankton feeders, and only about 25 out of the 350-plus species have ever been known to bite a person. The great white is one of them.

Lion

RRRRRRR

Lying on its back in the afternoon shade of an acacia tree, paws akimbo and relaxed, a lion in midafternoon may look like a stuffed animal or your friendly pet. It's not. Weighing 200 to 400 pounds, and able to leap 20 feet, it can break your neck with one bite. Lions eat as much as 60 pounds of meat at one sitting, and there are always several around tearing at the flesh together. Its roar is loud enough to be heard more than 5 miles away. Each lion has its own roar, too, roaring loud to keep track of its fellows and to deliniate territory or challenge another group's turf. You are advised to listen for it.

Lions hunt mainly at night, usually in just a few hours of intense activity out of every day. The females do most of the work in their stalking groups, the males doing their share only when the prey is especially big. Their favorite food is not humans but zebra, buffalo, wildebeest, any other large ungulate, though they will eat smaller animals, even porcupines, if they have no other choice. True carnivores.

A pride of lions is a fluid group. Its core is a few up to about twenty adult lionesses, mostly sisters and daughters or aunts and nieces, and their cubs. They not only hunt together and share the kill but watch over each other, even feeding each other's cubs; and they groom each other companionably when at rest in their more or less permanent territory. Male cubs leave the pride at about the age of three and roam with their brothers and cousins for a while. These groups of males, once mature and strong, then vie to take over various prides of female lions for mating.

Within a female pride or even a male-only group, lions are amazingly egalitarian. There is no top dog, alpha monkey, or chicken pecking order, although this is almost unheard of among animals. Aggression is controlled, probably because animals of this strength who did not control it among themselves would have driven themselves to extinction long ago. When a group of male lions takes over a pride, the animals don't fight among themselves over the females but adopt a principle of "whoever gets to that female first, gets her."

There is some violence among lions, of course. When a group of males moves in on a pride, the females who have other males in residence to help (and even sometimes those who don't), fight hard to repel them. But often the invaders win, evict the larger cubs, then kill all the smaller ones. Conveniently for the new males, the death of cubs throws females into estrus, the state of sexual receptivity, and, in this case, all at once. So the males are rewarded by being able to father a new set of cubs. They will stay with this pride and try to protect it for at least a couple of years, often much longer, until their own cubs are grown up. The males may also go out on the "marriage market" all over again, taking over an adjacent pride if they can, in addition to their own. Some of these battles over prides lead to deaths among the males.

Because there is greater conflict among the males, they seldom live as long as eleven years. Females can live sixteen to eighteen years. The dry season, with its dearth of prey animals, is much harder for them all to endure than the wet season. During true droughts, some prides even have to split up, and the lions roar less, preferring the quiet, comfortable shade of the acacia tree.

Tiger

PREDATOR FELINE

William Blake wrote, "Tiger! Tiger! burning bright/In the forest of the night,/What immortal hand or eye/Could frame thy fearful symmetry." In each of the recent decades, tigers have killed about 500 people per decade. And there really are "man-eaters," as they say. Tigers who enthusiastically hunt and kill people are usually the wounded, maimed, infirm ones that cannot hunt the usual prey. Manhunting may begin when the tiger pounces on cattle, only to observe certain even weaker creatures—people—trying to chase them away. These people are so easy to kill that a brand new hunting habit develops. Tigers—again just a few of them—have been doing this for at least hundreds of years. Female manhunters have even been known to teach the habit to their cubs. Tigers sometimes "clean up" battlefields or murder sites too. A person already dead is yet easier prey.

Since only a few tigers have "a taste for human flesh," as the expression goes, some conservation groups have propped up dummies (not real people) dressed in real people's clothes in the forest and rigged with just enough of a shock to make an attacking tiger think people have become unpleasantly electric. Another trick under investigation is to set out water for the tigers during droughts, the theory being that they enjoy humans partly for their bodily fluids.

In their jungle life, tigers are beautiful and beautifully camouflaged. Their coats glow like patches of sun in a clearing; their walk is powerful and quiet. They are usually resting in the jungle but are always alert even when they are drowsy, and they have excellent vision. Look for tiger evidence as high scratches on tree trunks; these marks are made when they sharpen and clean their claws and mark their territories (in addition to their scent-marks). Tigers have been stalking India at least since the third millennium B.C.

When hunting, usually at dawn and dusk, tigers patrol their territories and hang around the water holes. Then it is stalk, leap, pounce. In the water they are powerful swimmers, but they watch carefully for crocodiles, and it is said that they dislike getting their faces wet. In fact, when going in the water just to cool off, some have been observed going in backward.

Their level of aggression varies. The most aggressive sound is a woof and a cough, used with each other in fights. When hunting, they are absolutely silent. Mothers with cubs will even occasionally meow. When a mother is with her cubs (usually six or seven per litter), she will attack even a male tiger to protect them. Also beware when the male tigers are searching for a female and have enhanced aggression. When a male and a female are in their mating bouts (sex every twenty minutes for most of two or three days), they are the least dangerous.

A case can certainly be made for killing the "man-killers," if all else fails. But these are only a very, very few tigers. The rest of the graceful and awesome creatures should get our attentive protection. With only about seven thousand left in the world, they are now considered endangered because there has been too much tiger hunting, too much tiger fur wearing, too much tiger-bone using, and not enough habitat protection. There is one large sanctuary a couple of hundred miles south of Delhi, India, but many more are needed.

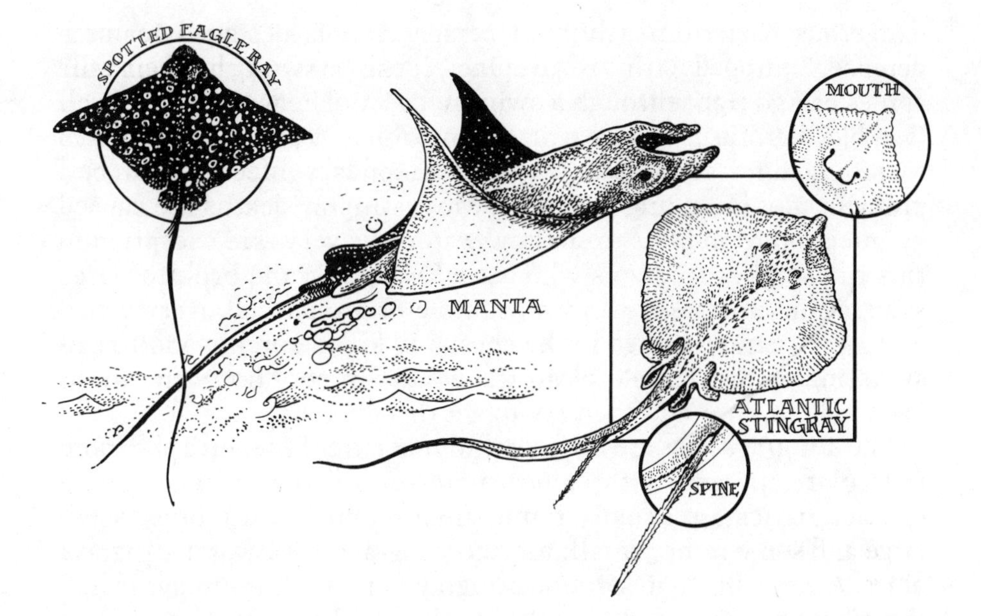

Stingray

COUSIN TO THE SHARKS

When walking into the ocean, it is wise to do the "stingray shuffle." Shuffling up the sand ahead of you with your feet gives nearby stingrays time to escape. Though most are shy creatures, they will sting when disturbed.

Stingrays typically rest on the bottom in warm ocean or estuary waters, flipping sand or mud over themselves with their "wings" to perfect their ambush. They usually enjoy dinners of worms and shellfish, but they will eat anything, even another stingray, if it moves along or near their low-down territory. To grab their food, they flip up quickly and then hunker down on top of it, smothering the prey and then eating it. (Their mouth is on their underside.) When they do leave the bottom and swim to the surface, they use a special veil that protects their eyes from the brightness.

Like the skates, they are cousins to the sharks, and all three have skeletons made of cartilage instead of bone. The stingray's tail, the dangerous part from a human perspective, is designed for self-defense. Equipped with barbed spines, it can be swung hard, and the spines can go right through a swimmer's leg. The wound will swell fast, bleed profusely, cause pain accompanied by cramps and nausea for several days, and it is slow to heal. Secondary infections can be a problem too, including gangrene. Most stingray deaths are caused by infection from bites in shallow water; the very rare exception to this rule applies to divers who actually land on the backs of large stingrays by mistake—if they stumble or are pushed over by currents, they can be spined in the chest and killed quickly. Victims of any stingray bite should clean the wound in very hot water for at least half an hour and then try to get to a doctor fast.

The stingrays that are the worst to meet are those with the more muscular tails and with spines more toward the tail end. These characteristics are usually found among four broad groups, some large and some rather small: marine stingrays, freshwater stingrays of the American tropics, round stingrays, and eagle stingrays. All of these have poison glands attached to their barbs. Both the eagle rays and the true eagle rays, similar and most dangerous, are quite large and have very strong teeth (used for cracking open clams and such, fortunately); they can weigh up to 1,000 pounds. All stingrays are flat with an eye on either side of their snout.

Two other rays are especially worth mentioning. The giant manta ray is not particularly dangerous but quite impressive. With a wingspan of 20 feet and a weight of 3,500 pounds, it can jump up high out of the water and come down fast with a scary splash. The whiptailed stingray uses its tail for more than self-defense; it whips fishes into pieces and then eats them up.

Beachcombers may have seen a stingray (at least in miniature) if they've come upon "mermaid purses" on the beach. These little envelopes hold the eggs of sharks or stingrays and, in the fifteen months it takes a little stingray to hatch, the purse sometimes gets washed up. These purses can be eaten, but beware of the adult stingray. Some of the more than thirty species are poisonous.

Wolf

HOW DANGEROUS?

The wolf has sunk its teeth into our imaginations. Its howl sounds the spirit of the wilderness, haunting and free. Then flood in the fears from reading fairy tales and werewolf stories. It is exciting, even exalting, to see a live wolf, its paws huge and its dignity unbounded.

Is the wolf really dangerous? It can tear chunks of flesh off a living deer, kill a moose by tearing through its 4 inches of fur and tough hide, even kill another wolf that tries to invade its territory out of desperation. It eats about 20 pounds of meat every day, or the equivalent of eighteen deer a year. And there are plenty of far-out folktales about it snatching human babies, treeing hunters, and setting upon adults along with its pack.

But wildlife researchers all agree that wolves are not usually dangerous to people. Wolves are shy and usually very difficult to see since they avoid people; they kill deer, elk, and moose instead.

There are, sadly, not enough wolves left to fear. Canada has only about 50,000 wolves spread through all its wilderness. Alaska has just 4,000 to 8,000. There are 1,500 to 1,700 in the northwoods of Minnesota, and about 100 in Michigan, Wisconsin, Montana, Wyoming, Idaho, and Washington state. Proposals have been made to reintroduce the gray wolf into Idaho and Yellowstone, and the red wolf has been reintroduced into North Carolina and the Great Smoky Mountains National Park, though livestock owners and hunters are still fighting these efforts. Opponents are even trying to get both the gray wolf and the red wolf off the endangered species list, since some of the red wolves tested seem to be hybrids of wolf and coyote. The Mexican wolf, whose range extends into the American Southwest, is almost extinct.

All wolves (except possibly the red wolf), including the timber wolf and the white arctic wolfe, are subspecies of the gray wolf. There are thirty-two subspecies worldwide; twenty-four of them were once in North America. Now that the wolf is classified as either endangered or protected in the forty-eight states (but not Canada), and the wolf slaughters that occurred up to the early twentieth century are over, wolf populations do seem to be expanding south and growing slightly.

The wolf needs meat, and that means wild territory. A pack of ten to twenty wolves requires 50 to 1,000 square miles in which to hunt, its boundaries marked with urine and no-wolf-lands between the territories of adjacent packs. Only when their wild food is gone do wolves bother livestock. Each pack includes one breeding pair (who are "about as monogamous as we are," one researcher says). Most of the others in the pack are their very young offspring, as well as subordinate animals who are their remaining offspring from previous years. Everyone helps to take care of the pups, by watching them when their mother is hunting and regurgitating food for them to eat. A lone wolf is usually a male but occasionally a female who would probably like to start or to take over another pack.

Wolves howl together in their wilderness for several reasons. One is simply for the social experience. Another is to find out where other pack members are if they have become separated. And another is to discover where strange wolves are and to defend their own territories from them (particularly if that includes a kill or young pups).

In addition to their magnificent howls, wolves convey messages to each other through their tail positions and through facial expressions such as the play face you can also see on your dog sometimes. They will also huddle around the alpha (leader) male, sometimes to all touch noses before going out to hunt together.

Most people know about the wolf of the woods but may not be familiar with the exquisite white wolf of the high Arctic tundra. One especially friendly pack was discovered by accident by a wolf researcher on Ellesmere Island near the North Pole, and he believes that this pack had never been harassed by human beings. These beautiful white wolves were curious, and the researcher played subordinate wolf by lying low and whining enough to reassure them. Soon the pack of seven approached him. And quite soon afterward, they felt comfortable enough to urinate around his belongings, then even leave him alone near the tiny pups when they went off to hunt musk oxen.

This pack and this researcher have more or less adopted each other, and the latter went back the next summer to find the pack had grown to eleven. The five new pups were unafraid and neither, of course, was the scientist. He is keeping their location secret.

Lamprey

BLOODSUCKER

Up north where we go in the summer, there is a story that I am almost certain is true even though it is quite old. A local woman, whose name was well known in the town and who had always been a strong swimmer, wanted to swim across the bay, a blue swatch of Lake Michigan about 4 1/2 miles wide. She was smart enough to have a boat accompany her and was doing fine for the first couple of miles. Then, out in very deep water and with her legs chilled to about the temperature of the local fish, she was attacked.

Her predator—a lamprey—attached itself to her leg with its suction mouth and began to suck her blood. Fortunately, her friends in the boat were able to beat it off quickly. She lived to swim again, bravely. But not that day.

This lamprey had confused her with a fish, perhaps even one of its favorite local delicacies, whitefish, as the town's residents then agreed. And, even now, all over town, brothers and sisters and cousins are still scaring each other with this story. Even though I would indeed like to discourage tourism to this beautiful bay, I must confess that people also agreed that had she not been out so deep and had her legs not chilled so thoroughly, the event would almost certainly never have happened.

Lampreys, which look like eels but are not (they have three fins), slither through the oceans and freshwaters of the Northern Hemisphere. They accept a wide range of water temperatures and salinities, a flexibility that has hastened their spread.

They are living examples of very old vertebrates, the jawless fish of Silurian times 300 to 400 million years ago. Lampreys not only lack hinged jaws but they are also missing ribs, a thymus, lymph vessels, and a genital duct, and so are very primitive indeed. Their first confirmed fossil ancester (progenitor of all the lampreys and probably all the hagfish of the world) was discovered as recently as 1991; it had longer fins (really tentacles) and more prominent eyes than the lamprey of today.

Of the forty-one species of lamprey known, eighteen are parasitic. These suck the blood and chew the muscle of prey such as mackerel, cod, salmon, herring, hake, swordfish, and even chomp sturgeon and basking sharks. They have been known to attack sperm whales and, very rarely, land animals such as humans. Lampreys use their disk-shaped suction mouths to attach to the prey, then bite through scales and skin with their horny teeth. An anticoagulant is secreted by their mouths so that the bloody dinner will flow freely. The prey dies from blood loss or from the fungi and bacteria that grow as secondary infections on the wound. But by that time the lamprey has left to suck out another fish.

They have a quite strange life cycle. It begins in a stream near the ocean or lake they will cruise as adults. The lamprey father uses his mouth to excavate a 6-inch nest in the gravelly bottom and the lamprey mother releases 60,000 to 240,000 eggs, each just about 1 millimeter long. The parents then die, and the young hatch eight to twenty days later. Blind and toothless, they live in their home stream for three to six years, filtering plankton and detritus for food, hiding safely in the little tubes they have burrowed into the streambed. Once they reach 10 to 12 centimeters long, they finally migrate out to the ocean or lake, acquire vision, become quickly carnivorous, and grow fast. From a brownish white, they are transformed to a blue-green with a silvery belly. They are slimy, like the hagfish to which they seem to be related. This change of shape and feeding behavior is curiously amphibian, scientists note.

The species called the American sea lamprey, the one in my scary swimming story, came in from the Atlantic to the Great Lakes by attaching itself to the bottoms of ships that came from the ocean into the St. Lawrence River, then farther, via the newly cut Welland Canal. Great Lakes fisheries suffered for many years, but this lamprey's numbers have been controlled now—and readers who want to do some long-distance swimming there are probably quite safe. Just don't dress up like a fish.

Tarantula

CREEPY, BUT NOT TOO DANGEROUS

This is a musical creature, in a way. It has inspired a dance called the tarantella. Thereby hangs a tale, which begins in Taranto, Italy, on a day long past. Someone in the town was bitten suddenly, it is said, by a tarantula and that unfortunate person sank into a deep lethargy. The victim could be aroused from it only by music and frenzied dancing. Dancing to the point of collapse—to work the giant spider's venom out of the body—seemed to be the "cure." As time went on, others thought they were bitten too and, in turn, also imagined that the dance should be danced on the anniversary of the original tarantula's bite.

All of this is now considered to be a manifestation of mass hysteria, and some people do not even believe that tarantulas ever lived around Taranto, Italy. Neither does exercise remove venom from one's system. But it is a nice story, a nice dance, and nicest of all, perhaps is it to have any arachnid honored.

Almost all species of spiders are venomous and bite to immobilize their prey, but most species are so small that the amount of venom that kills insects for them is trivial to us. The venom of tarantulas is indeed neurotoxic, causing spasms in heart, lung, and other muscle tissue, even paralysis. But it usually has very little effect on people and, in fact, these spiders rarely bite even when handled. Some people even consider them sweetly furry to the touch.

They are probably a bit big to be considered cute, though. North American tarantulas have a leg span of about 5 inches, and there is an Amazonian one that is 10 inches across and can kill birds. The many species of this spider live in both deserts and tropical areas. One tarantula species, found all over South America, lives alone in a burrow for up to twenty years, biting and then eating insects and small reptiles, rodents, and frogs. It comes out only to find a mate, which quest can lead it into people's houses. Once safely found and mated, the female tarantula becomes a solicitous parent; she guards the egg case and even watches the baby spiders until they can get their own food.

Spiders in general deserve not only our respect but our admiration. Spider silk is so wonderfully strong that it can hold five times more weight than a similarly slender piece of steel. It is also amazingly elastic, able to stretch one-fifth of its length without breaking, which gives it a tensile strength greater than that of steel. Their silk is at its strongest if the spider draws it out from the spinnerets more quickly rather than more slowly. These qualities have not been lost on one English scientist; he is trying to use genetic engineering to mass produce it as a "biosilk."

Spiders make several kinds of silk, and the strongest is the variant used in their draglines (they spin out and then drag this silk behind them when they leave their home areas, even to become airborne). Other kinds are used for their egg cases, nursery areas for their young, and, of course, webs or lined burrows, depending on the type of spider. Perhaps the most unusual recent finding of all is that many spider webs reflect ultraviolet light. This is a silken attractant to insects. Come hither, said the spider to the fly.

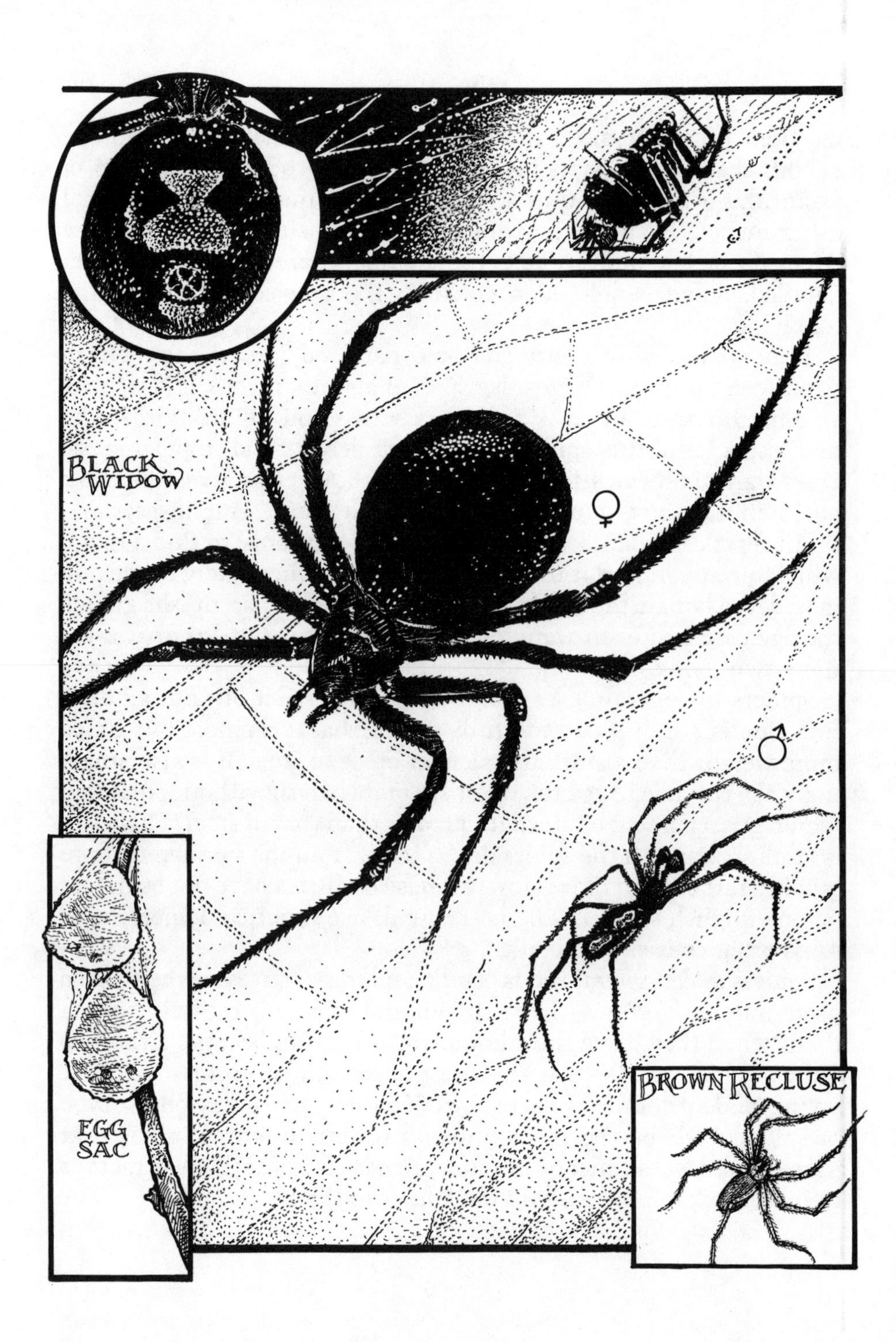
BLACK WIDOW
♀
♂
EGG SAC
BROWN RECLUSE

Black-Widow Spider

DEADLIER THAN THE TARANTULA

Drop for drop, this creature's venom is more powerful than a rattlesnake's. Draping its victim in silken thread, it empties two venom glands into the body through its fangs. Once the prey has been killed by the black widow's neurotoxic venom, the spider wraps it up in more silk, then waits while the venom dissolves the prey's tissues from within. The victim is generally an insect.

Since this spider lives in all forty-eight of the contiguous United States, as well as in most of the rest of the world, its bite's effect on humans is worth noting. First, this bite causes almost no pain. Later, though, intense pain will mark the spot, extending also to the groin and, to a lesser extent, the legs. The abdominal muscles and other large muscles grow increasingly rigid. Sometimes the symptoms are accompanied by nausea, headache, profuse sweating, and difficult breathing, even shock. All this goes on for two or three days, by the end of which most people have recovered. However, about 4 percent encounter lung paralysis—and die.

In one study, 88 percent of the people bitten by black widows were men using outhouses. You can guess where they were bitten. This spider also likes to hang out around barns, garages, and houses. And they bite a lot compared to other spiders; in another study sixty-three out of sixty-five spider bites brought to medical attention were from black widows.

The black widow with the business bite is the female. Males have poison glands too, but theirs are much smaller and become inactive by the time of maturity. So here is how to identify the sexes: the females are glossy black or sepia, with a red hourglass shape—or stripe or blotch—on their undersides; the males are near-black to a mottled brown and feature a pair of knobs on their heads. The females hang upside down on their irregular webs, waiting. Both sexes are shy, and they definitely do not like to be disturbed.

Another spider, who rivals the black widow in venom, is also found in the United States, especially in the West and Midwest. Called the brown recluse because it likes to stay in corners or under objects, it also delivers a bite that hardly hurts at all to begin with.

Then symptoms crescendo through pain to lesions that soon form open wounds that can take weeks to heal. These tiny spiders can easily kill a child.

Looking on the brighter side, spiders have not only long fascinated people but are said to have inspired several inventions: the suspension bridge, the diving bell, and even the flying kite. Probably a lot of nightmares too.

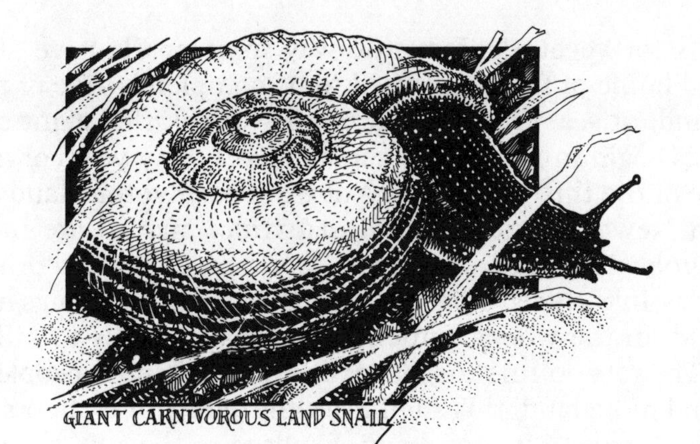

Land Snail

GIANT CARNIVORE

Don't roam around in the woods of New Zealand at night if you don't want to feel something creepy climb up your foot. It could be the giant carnivorous land snail, its gold and amber shell holding a black or striped creature a full 8 to 12 inches long. Giant, indeed, for a snail. It is hunting there, especially in its favorite damp weather, and looking to eat foot-long bush worms and other (slow) prey. They don't eat people but would certainly feel very unpleasant.

The latest word on this creature is that nearly 1,000 of them were illegally brought to this country—not by a theme park owner with a twisted mind but by an animal broker seeking to sell the giant snails to pet stores. The snails did indeed get into pet stores, next into the terrariums of homeowners. Unfortunately, they are also a threat now to U.S. crops. These snails eat almost anything, including fruits and vegetables. They tuck away about one-fourth of their body's weight in food each day. And they can have about 1,000 babies in a lifetime. Snail owners are urged to turn them in to their nearest U.S. Department of Agriculture office.

Usually snails are slimy, but this one isn't. It is a rather primitive creature, but one so successful over its millions of years that it never seemed to need to evolve slime as a protective coating, the way the more modern snails did. It also never needed to abandon its carnivorousness for vegetarianism, the way most snails have. The New Zealand home of this snail used to be completely free of predators from land, or sea, or air. When not disturbed, this giant carnivore lays eggs beginning at age fifteen and lives forty years or more.

Some of the thirty-six species of giant carnivorous land snails, at least in New Zealand, are almost extinct and all are in trouble. Their problems began when pigs, deer, goats, rats, and other mammals were introduced onto the islands. Some of these pigs have gone wild and, in the woods, puncture the snails' shells to kill and eat them. The rats do this too. Also, as more and more people develop more and more land, this snail loses the damp forest floors where it needs to burrow 6 inches down. Exposed to the sun, the soil dries out, then the snail—its slimeless body without protection—dries out too. Then it dies. Laws are now in place to protect it and to forbid the collecting of its shells, which are beautiful enough to have created a black market. It is also illegal for them to be exported to other countries, since natural controls on their population might well not be in place.

Africa has giant land snails too, which are actually thriving. They also inhabit places like Hawaii and New Guinea where they were introduced to provide an extra food source for the human inhabitants. There are now, in fact, too many of them and they are crawling right into people's homes as well as becoming a major agricultural pest. Though this one, unlike New Zealand's, is not a carnivore, both illustrate the same lesson about humans introducing new species to places: *don't.*

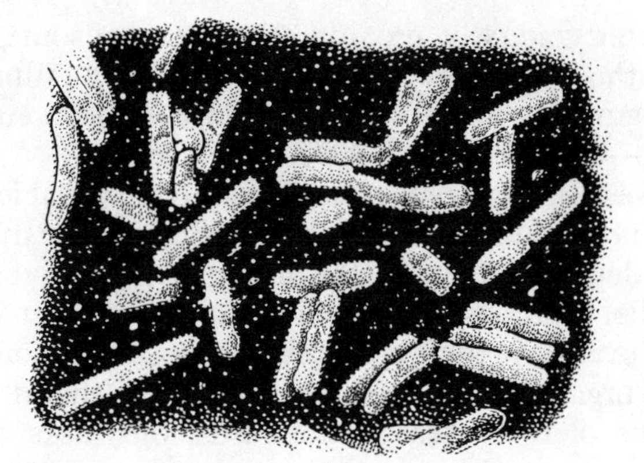

Legionella Pneumophila

SMALL AND SNEAKY

This "creature" achieved its Warholian 15 minutes of fame in the Bellevue Stratford hotel in Philadelphia in 1976. Legionnaires' disease sickened 221 people and killed 34. And no one knew why. This organism probably also caused Pontiac fever, an outbreak in 1968 in Pontiac, Michigan, where a much greater percentage of a building's workers caught the disease but found it milder, shorter, and never fatal. It might also be the "criminal" behind an outbreak in 1965 in the Washington, D.C., area and one at a Spanish resort in 1973. In all these cases the organism was isolated, but the symptoms were not exactly the same.

Legionnaires' disease creates symptoms in people that begin two to ten days after exposure. The victim feels generally sick, especially with muscle aches and headache. Then the fever spikes up, accompanied by a cough, chest and stomach pain, diarrhea, then difficulty breathing. The fever hovers at 102 to 105 degrees Fahrenheit, with raspy breathing.

By this time, most of the people who have encountered this organism have checked into a hospital. They usually have pneumonia, too, and feel even worse for several days. Some even feel a strange confusion. If the proper treatment is not administered, 20 percent will die of severe pneumonia or shock; the other 80 percent will feel better gradually, provided they receive some mechanical help to breathe and to prevent complete kidney failure. But even after this regimen, some survivors feel week for months and may have partial—and permanent—lung damage.

What causes all this havoc is not a readily observable, combatant creature; it is not even a typical pneumonia organism, since true pneumonia does not include this dimension of diarrhea or confusion but rather includes a runny nose and sore throat which is not found here. In fact, it took epidemiologists many months to actually see this organism and even longer to prove that it was a bacterium. As one of the researchers pointed out, it is an organism "fastidious in its nutritional requirements and . . . hard to grow in the laboratory . . . whose precise ecological niche is difficult to discern."

More study revealed that it lives in air exchange condensers where it probably lodges after flying into the air on bits of soil. Though its favorite habitat is the soil, it can survive not only on a damp air condenser but even in ordinary tap water for more than a year, longer on partly dry equipment.

It is now suspected that out of 2.4 million pneumonia cases in the United States each year, perhaps 7,000 to 36,000 are caused by this organism. The best antibiotic seems to be erythromycin, though it is not always effective. Victims are often the middle-aged or elderly; cigarette smokers or heavy drinkers of any age; small children; people living, working, or staying near a construction site (where it becomes more easily airborne from the soil); and those with some preexisting medical problem. It seems to enjoy its greatest level of activity in summertime, all over the world. *Legionella pneumophilia's* numbers are, it seems, legion.

Killer Whale

CALL IT *ORCA*

It is not even a whale and belongs instead to the dolphin family. It is definitely a killer, though of seals and sea lions, and it can flip one of these bulky creatures into the air for fun before chomping it dead. It is certainly a killer, too, of salmon, tuna, other large fish, and squids, and can even herd them into an inlet where they meet their death. It is surely a killer of fellow dolphins too, sometimes leaping up to catch them in the air at the top of their own leaps. It is also an attacker of whales as large as the blue—a group of killer whales will force a blue whale's mouth open to tear out its tasty tongue. Killer whales usually swallow their meals whole or in very large chunks.

People should call it *orca*, not *killer whale*, though, since its danger to humans is as small as its body is big: if you try to hurt an orca, it will probably try to hurt you back, and that is about the only time you need fear it (unless you get in its way and it hurts you by mistake). Those stories about murderous orcas apply only to a few in captivity. This creature just wants to mind its own business in the sea.

The orca is huge. Males can be 30 to 33 feet long and weigh 9 tons, and the females are slightly smaller. Both have a black dorsal fin straight up from the middle of their backs, and both are dazzling in black and white, each with slightly different markings.

The animals live in pods, social groups of up to 200 individuals, composed either of adult males, females and their young, both sexes, or young males alone. The mixed groups often stay together for generations. Occasionally, several pods get together too. A mother orca has her first baby when she is eight or nine years old and gets help taking care of it from another female orca in her pod. She has a calf every three to eight years after that, for most of her life of thirty to fifty or more years.

These are very playful animals and have been seen throwing kelp into the air like balloons, splashing with their tails, "spyhopping" to see what's going on, and doing acrobatics of many kinds. Their sociability is also displayed in a communication system yet to be completely decoded. It includes a whistle-honk that seems to be a

signal to converge on a school of fish and many other clicks and whistles and squeals. One pod's dialect is slightly different from another's. Each whale also adds a signature sound to its communications that distinguishes it from every other orca. When two pods greet each other, they sometimes line up facing each other, then move in between each other's lineups to touch.

Hunting methods are quite sophisticated pack maneuvers, which have earned orcas the nickname "sea wolf." Using their sonar, they locate a school of prey fish, then push them into the nearest inlet to feast. To rest, whales in a pod assemble near the surface close together, then go up to breathe rhythmically at the same time. These rest periods of one to six hours a day seem to be their only "sleep" away from their hunting and playing lives.

Orcas travel about 100 miles on a typical day and can live in all the world's oceans and some rivers; they are most common at high latitudes. They are usually found within a couple of hundred miles of shore in the United States, including the whole California coast and, in the East, from New Jersey northward.

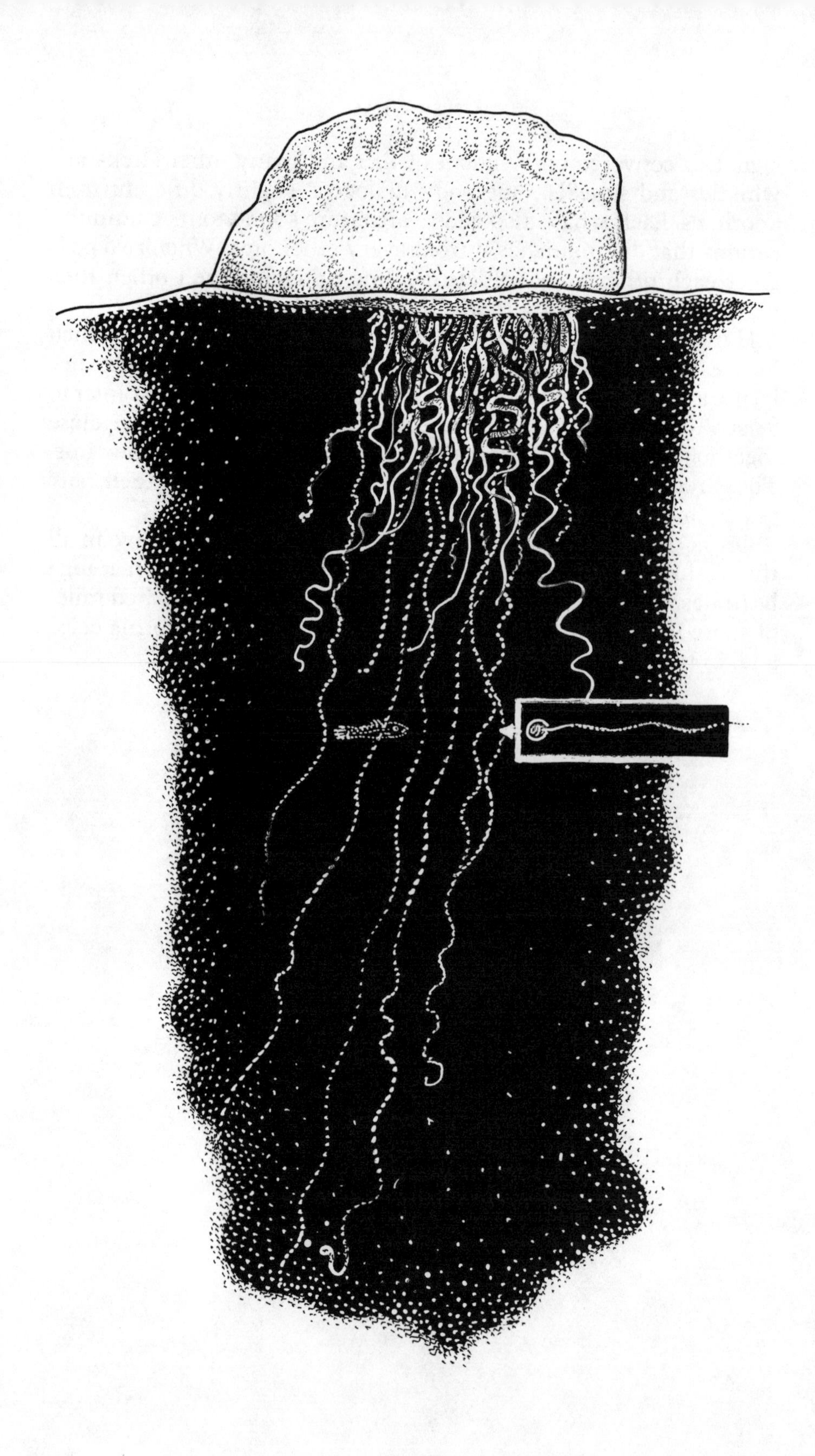

Portuguese Man-O-War

SLIMY STINGER

With its tentacles fully stretched for feeding, this can be one fierce 200-foot-long jellyfish. With them retracted, it is about 3 feet of innocent-looking float, a blue bubble dancing on the sea. In either case, a full-grown Portuguese man-o-war has twenty or thirty tentacles, and each packs up to 750,000 cells for stinging. People hit by enough of these little harpooned stingers can go into shock, lapse unconscious, feel ill for several days, be decorated with welts for several weeks more, and then carry scars for the rest of their lives. These creatures can sting even when washed up on the beach, long after they look dead.

The man-o-war, named by medieval sailors for a Portuguese sailing ship, indeed rides the wind, its bubble above the surface and tentacles beneath. It usually sails at a 45-degree angle to the wind but can trim its "sail" to a certain extent to change this angle; also, some are right-sided and some left-sided in the placement of their bubbles.

This jellyfish, about 12 inches across the bubble at maximum size, is found in warm waters all over the world. The hugest concentrations float in the Sargasso Sea in the mid-Atlantic. Since they cannot truly swim, a storm or strong currents can wash up thousands onto beaches anywhere in their territory, even a small stretch of Florida or southeastern American beach.

Not truly a single animal, the man-o-war is a colony of up to 1,000 polyps, of four different types, acting together. The float (or blue balloon) is a single overgrown polyp. The tentacles are made of many polyps of a second kind. Then there are the stomach polyps, creating for it many stomachs that can all eat at once if called upon to do so; in fact, the tentacles are such an efficient fishing skein that one observer watched a large man-o-war that had caught twenty little fishes and was eating all of them simultaneously. Its fourth kind of polyp makes up the reproductive organs. There are no organs for elimination, and wastes come right back out of its mouth and into the sea. The creature has no brain, no heart, and no blood vessels.

Enemies of the Portuguese man-o-war are the ocean sunfish, one kind of sea slug, violet sea snails, two kinds of turtles, and one kind of octopus that breaks off its prey's tentacles and then uses them to sting other creatures to get the rest of its dinner. Do not get in the way of this jellyfish.

Moose

WATCH OUT FOR THOSE ANTLERS AND HOOVES

The moose is the largest member of the deer family alive and may be the largest mammal with antlers ever. To see one is to feel fear along with awe. Usually, the fear is not warranted, since the 50–60 pounds of food it eats a day never include people—it relishes lichen, moss, tree barks and shoots (of both conifers and hardwoods), fresh new grass and mushrooms, and water plants from water lilies to algae to wild rice.

Nonetheless, make sure not to disturb this giant ungulate when the cow moose, the one without the antlers, is with her calf. During its first 1 1/2 years of life she is very protective of it. The younger the calf, the more protective the cow moose, since a newborn moose will follow anything and the cow does not want it to follow you. Until she realizes that you are no threat, she will probably charge if you approach. A mother moose is a mother most of her life too, so this means watch out for mother moose most of the time.

Also beware when the male moose, the one with the antlers, is in rut. This rutting season, when its body is rich in testosterone and it is searching for females, is generally September to October. He may be having an antler battle with another male or shoving away another animal, then turn to charge a hiker. The best thing to do is climb a tree, fast and high—moose can run at 30 to 35 miles per hour and have even been known to charge helicopters.

The antlers are not the main danger to people, however. They are not even always on males (and never on females); moose shed their antlers every winter, growing a new pair gradually in the spring until the set may weigh 60 to 85 pounds. Each moose's antler shape is a bit different, and they also change shape throughout a male moose's life.

The real problem is the hooves. Moose of both sexes can use them like street toughs to keep even a pack of wolves at bay (wolves almost always get only weak moose). With a 600 to 1,500 pound sure-footed animal behind them, these slashing hooves can kill you. Its threat gesture is a lowered head, flattened ears, raised hair, and flaring nostrils.

They are generally minding their own business, though sometimes noisily. Moose peacefully browse the north woods, including its open areas and marshes, feeding mostly at dawn and dusk. They stand in bushes nibbling, go down on their knees for grass, swim for water plants, even dive as deep as 18 feet down for vegetation. Their name, from the Algonquin, means "twig-eater."

In fall, you may hear a male scraping his antlers against a tree to get the "velvet" off. In spring, you might hear a female grunt to her calf, or, if alarmed, make a barking sound; and a male might be (literally) moaning for a female or attacking shrubs or small trees in his excitement. During the middle of the day, they are almost always completely silent. Usually just a cow and calf (occasionally twins) are seen together, the male along only rarely. If you see a few males together, it is probably the beginning or end of the rutting season, or food has attracted them to the same spot.

Their necks are long, they have a very muscular and floppy upper lip called a muffle for pulling at plants, and there is a handful of skin hanging down from their necks called a bell. Their legs look overly long and their vision is poor. An adult is 8 to 10 feet long and 5 to 7½ feet tall to the shoulder. Another awkwardness: if a cow is a little upset, she will squeeze her legs together and urinate on her feet.

Besides those hooves and antlers, they do cause other problems, in Alaska, at least, since they are so numerous there. They stray into people's gardens and eat all the vegetables and ornamental shrubs. If people feed them more directly, they start coming up and demanding more. They often won't get off the trails or even the highways, standing their ground with assurance. Be assured that in a collision with a moose, no one wins.

Centipede

BITING PEST

In probably the only poem ever composed about centipedes, an anonymous author wrote,

> A centipede was happy quite,
> until a frog in fun
> said, "Pray which leg comes after which?"
> This raised her mind to such a pitch,
> She lay distracted in a ditch
> considering how to run.

Centipedes, perhaps not this cute, may have between 15 and 100 pairs of these confusing legs. Yet they do walk gracefully. The smaller species scramble throughout eastern North America and elsewhere, either outdoors under rocks and logs and in the leaf litter, or inside near drains and sinks. They are incredibly common—one naturalist who counted the insects in just 1 acre of English pasture found centipedes and millipedes numbering 38 million creatures. When inside (more rarely), they oh-so-helpfully eat up cockroaches, houseflies, and other insects, growing only to 2 inches long themselves and coming in soft grayish browns. They prefer crevices and have flattened bodies.

Both in and out, centipedes grab their prey with a set of claw-like and poisonous jaws. They also brew nasty-smelling and bad-tasting chemicals in their bodies to keep away their own predators. If handled, they can bite painfully (but nowhere near fatally). If you find the local house centipede really creepy (as I do), ponder the large tropical species. At 10 to 12 inches long, this one kills and eats mice and small birds.

It may seem strange, but centipedes are not insects at all (they not only have more legs than the latter do but theirs are attached differently). And they are among the most ancient land creatures—the latest fossil discoveries put them at 414 million years old. When we've lasted that long, we can feel superior.

Sea Wasp

KILLER JELLYFISH

It may seem surprising, from our limited perspective as land creatures, but there are more venomous things in the sea than there are on dry land. Almost all jellyfish have stingers, but the best armed of the sea wasps can actually kill a person—and in just about five minutes. They may be the most dangerous creatures in the sea. In fact, at one beach in Brisbane, Australia, where records were kept for many years, sea wasps killed five times as many people as sharks did. Even the turtles who eat these wasps become poisonous themselves and you might want to be careful sipping turtle soup.

Sea wasps are about the size of cabbages at their very biggest and, tentacles extended, the largest ones can stretch to 20 feet long. Each tentacle has hundreds of thousands of tiny stingers equipped with toxin. Unlike some of the other jellyfish, these sea wasps have tentacles grouped at the "corners" of their somewhat boxy bodies.

The creatures are wholly transparent. Although they can propel themselves through the water at about 6 yards a minute hoping to come upon food, they don't actually come after humans to attack. The idea is to avoid getting in their way, especially in the tropical waters where they all live. The Australian species are the most hazardous.

All jellyfish are amazing bits of architecture. Usually about 95 percent liquid, each comes in pale shades of blue, red, or purple, as well as white (transparent). All have tentacles and an "umbrella," and most, but not all, move by contracting their tops to force the sea water out, then by opening to let it back in again for the next stage of jet propulsion. To touch one (a small one is safest) is to know that they are firm, with connective fibers to hold their "jelly" together. When jellyfish are young, they are very different, only tiny polyps on the bottom of the sea. There, they look like the coral and the anemone to which they are related.

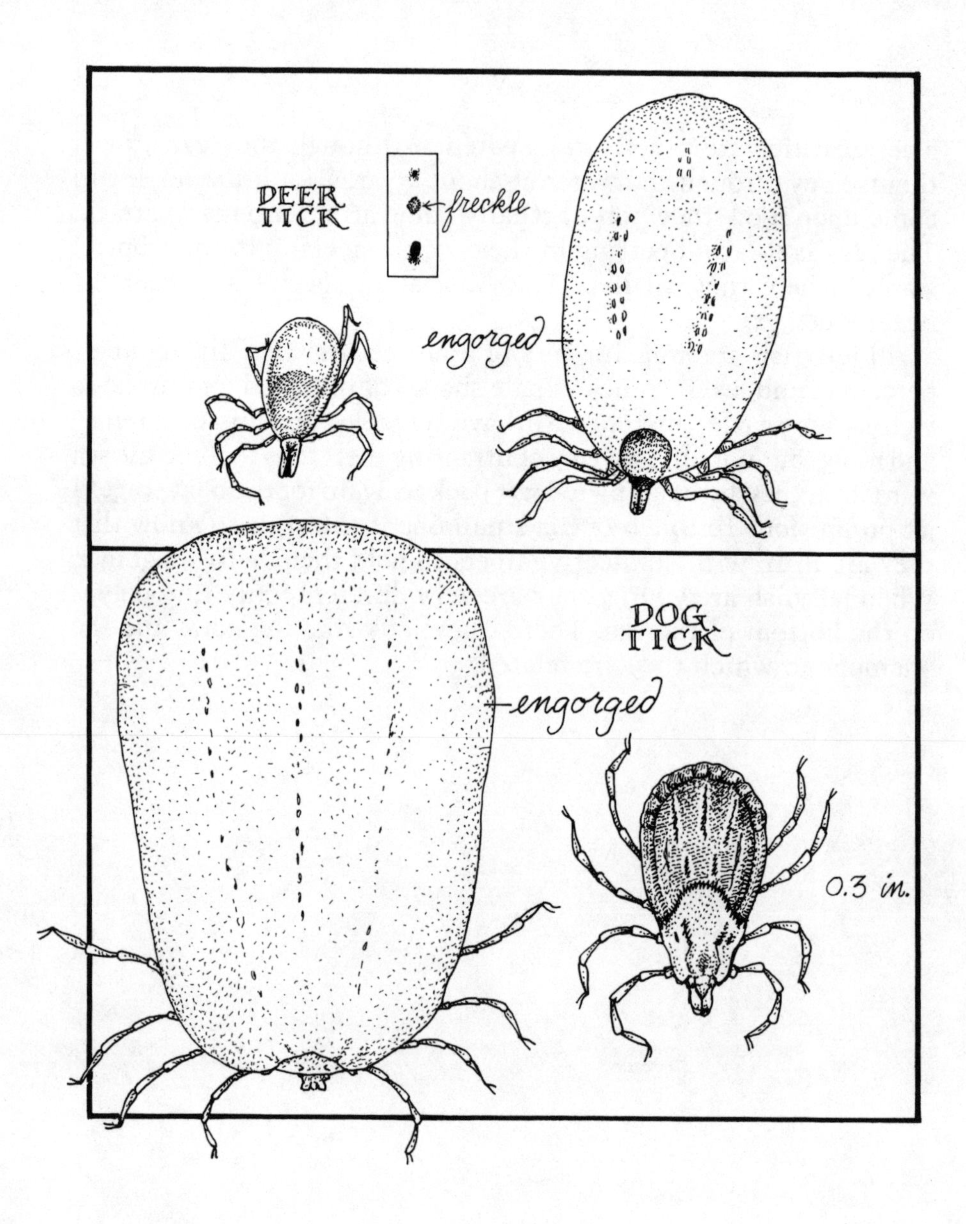
DEER TICK
freckle
engorged
DOG TICK
engorged
0.3 in.

Tick

UNWELCOME HITCHHIKER

They can't fly or even jump—just wait, and then scuttle—but ticks are experts at carrying diseases. What they bring along depends on whose blood they sucked for their last meal. The master list includes seven kinds of fevers, five varieties of encephalitis (swelling of the brain), seven other diseases mostly named for the places where they were first identified, and an additional ailment that is the most famous of all: Lyme disease. Some of these diseases are merely unpleasant, some fatal.

The best way to identify a tick is indeed to see it scuttle. They can be as tiny as a speck of dirt (until they have had a blood meal, whereupon they swell to about ten times that size). They have especially tough exoskeletons, making them usually impossible to crush between two fingers. Members of the arachnid family (along with spiders, scorpions, and mites), they have eight legs, though very tiny ones.

The life of a typical tick is two years long and is characterized by both patience and surfeit. Each tick begins as an egg in the springtime, hatches to a larva a month later, and then latches onto something for its first two-day feast, usually the first mouse that strolls by its home on the top of a blade of grass. The many offspring from one mother all lie in one area, so that any mouse coming along is very likely to become infected and the tick to get its blood meal. Then all fall and winter the full little tick rests. The next spring the tick larva becomes a nymph and attaches itself in this form to the next warm-blooded creature that happens by; this meal is even longer, about three to four days, and the tick sucks it from a mouse again or a larger creature such as a dog or a person sitting in the grass. If the prey happens to be a migratory bird, the tick will be in for an airborne adventure, which is how the ticks and the diseases they carry spread throughout the world. A shorter ride brings the tick home attached to your dog, passing the tick from fur to hand.

At the end of this most action-packed summer, the tick has become an adult and takes up residence even higher in the underbrush. This makes it easier to find a deer or a hiker who is hiking rather than sitting. And right on that new host, mature male and female ticks find each other to mate. The male dies afterward, and the female overwinters and then dies as her eggs start the whole wonderful life process over again in the spring grass.

But for a creature that is nearly immobile (until it finds something nearby to scuttle onto for the ride), life is not always this easy. The tick, ever looking for blood, must find it by smelling carbon dioxide (exhaled by all warm-blooded creatures) or butyric acid (exiting from some animals' skin). But what if nothing smells this way, or nothing comes by? Well, it waits. And it can remain in a waiting mode for *several years*, thus extending its life.

If a tick bites you, you will not even feel it; the tick exudes an anesthetic, presumably so that it can feed without alerting the host to its presence. Its little head buries into your skin, attached tightly with its own tough brand of glue-like spit, and it begins to feed. An anticoagulant from its body keeps your blood conveniently flowing, as you exchange microbes and bacteria with this tick.

In its bodily fluids will be one type or another of the tiny coiled troublemakers called spirochetes. They are small enough to pass through many of the filters the body uses to keep things out. Let us look at the effect of the ones that bring Lyme disease as an example. The first stage of this disease is the circular rash, beginning two to thirty days after the tick's visit. Fatigue, fever and chills, as well as terrible headache, neckache, and backache often accompany it. The problem is that some people escape all of these symptoms. Second-stage symptoms, which can also be masked, include muscle pain and meningitis, also sometimes skin sensitivity, heart palpitations, and dizziness. If you experience the latter, you may need to be fitted with a heart pacemaker temporarily. The last stage is arthritis attacks. In a few patients, this denouement includes extreme sleepiness, memory problems, and mood changes. The disease does not let loose any easier than the tick did.

There are ways to avoid this scenario. Inspect yourself and have a friend inspect you for ticks when returning from out-of-doors in the spring or summertime. All stages of the tick's life can infect you, so check carefully. If you see a tick, remove it with tweezers grabbing its head (or smother it with nail polish or oil, as a last resort). Put a little antiseptic on the wound. Watch for symptoms of Lyme disease or other fever and encephalitis symptoms; if any occur, hike straight to your doctor. A new vaccine against the disease, under development now, has worked on laboratory strains of the disease, on mice, and also on dogs. Whether it can work against the attack of the wild and wily tick on humans, only time will tell.

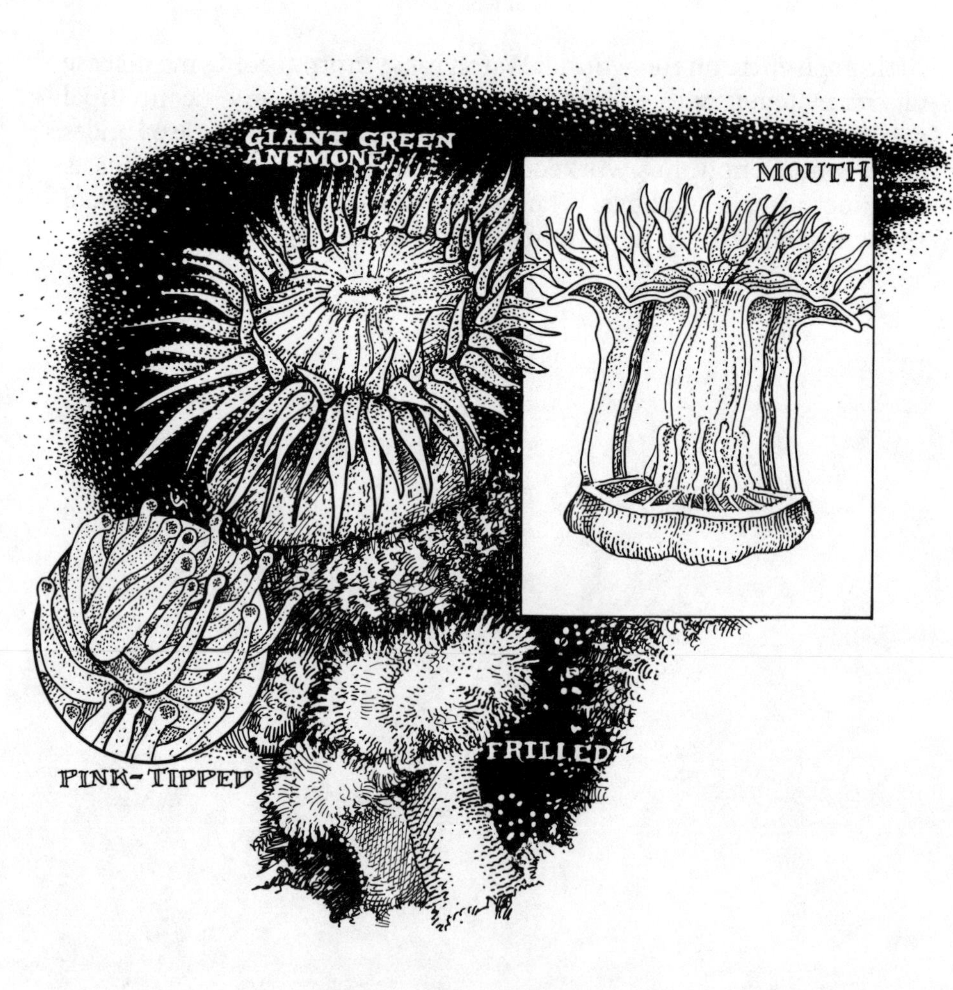

Sea Anemone

STRANGE CARNIVORE

Related to both jellyfish and corals, they come with beautiful names like rosy anemone, color-changing trumpet, dahlia anemone, and great green anemone. They "bloom," too, in all colors from gold to purple to pink to orange. But while they have the faces of flowers, they have the mouths of animals.

Sea anemones, are, in fact, carnivores who fire toxic weapons at their food. They also have good senses of hearing (by detecting vibrations) and smell, as well as touch; they can move around and are fiercely territorial. The almost 1,000 species of sea anemones sway colorfully in a broad kingdom: through all of the world's oceans from the Arctic to the tropics, and from shallow water down to depths of 30,000 feet.

To humans, their sting usually means some itching and swelling, though some species can cause fever, chills, stomach pain, and vomiting; and at least one species is poisonous to eat. In some cases, effects can last for several months.

But their weaponry is most typically directed toward food, not people. Anemones eat everything from shrimp and crabs to zooplankton and fish, as well as barnacles and mussels unstuck and washed their way by the tides. A large one was even observed consuming a small shark. They catch their prey with sticky threads fired out by cells called spirocysts on their tentacles. The silk wraps around a mollusk or crustacean's shell. They can also fire "harpoons" from nematocyst cells, also on the tentacles. These poisonous weapons are launched by the anemone at more than 10,000 times the acceleration felt by astronauts at lift-off.

Most people know that anemones are sensitive to touch, their tentacles grasping at whatever comes within their swishing arms. But they can also use an exquisite sense of the vibrations made in water by their prey, as well as smell, enabling them to blast out their poisonous harpoons properly. These active senses allow for excellent fishing even though anemones are blind.

Most anemone activity involves catching food with their tentacles. As every tide rises, they unfold and inflate them to become twice or more their resting size. They then pump water into their mouths, enough to keep their bodies full-sized even with the buffeting of waves and tidal currents. Though they are indeed cousins to the corals, they have evolved to lose their hard skeletons, giving them all the advantages of mobility. Their stalk (mostly hollow and used for digestion) and its base can glide along at about the rate of a centimeter per hour, and some young ones can even detach themselves and swim away. Anemones can thereby gain a position where the tides will sweep more food toward them, get out of the way of a larger, unrelated anemone, and avoid predators such as the starfish. At least one species even picks up its stalk to copulate, base against base, though most anemones reproduce by merely releasing eggs or sperms to the currents or cloning themselves (by dividing themselves in half vertically or horizontally).

Their territorial movements are amazing. An anemone can recognize its clones (genetically identical fellows) and will not fight with them for space. Some species will battle with strangers, though, inflating little sacs below their tentacle array. The tips of these sacs can then attach to the unwanted neighbor and release chemicals to hurt or kill it. In smaller amounts, the chemical makes the neighbor wise up and move away a bit. In a full-fledged fight, two anemones will move their tentacles completely out of the way and swing these sacs repeatedly at each other. They seem to distinguish relative from stranger in a way that resembles the operation of the human immune system—by matching the configuration of the antigens on the cell's surface to their own.

Perhaps even odder, if a species of anemone lives in a colony, the individuals who live near its edges develop more and stronger sacs for battle than do the anemones in midcluster. This has reminded researchers of the "soldier castes" of social insects. Some anemones lack these fighting sacs, and so use their tentacles for wars instead. In these cases, the tips of the regular tentacles have developed killer toxins similar to those in the fighting sacs, and different from the toxins used by the tentacles for getting food.

As the creatures become accustomed to their unrelated neighbors, they quarrel with them less. One species will fight only with anemones of its same sex. And others choose, instead, to pick fights with the corals on which they live, seemingly to prevent the coral from growing over them. Neptune only knows how long all these prize fights would go on if there weren't plenty of coral for anemones to live on. Usually they move out of each other's way after only a couple of rounds.

Beyond tentacles and sacs, anenome anatomy is very simple. They have merely a slit mouth and a throat, and release wastes up again through their mouths. The rest of the body is stomach, subdivided into many sections by thin-walled partitions. In times of plenty they can eat fast and grow splendidly; with scarcity, they shrivel up to tiny little things. They have no heart, no central nervous system, and no brain, not even a real collection of nerve cells beyond the basics for coordinating their behavior.

Anemones are not quite all fight and food. They get along very well with clownfish. These beautifully brilliant little fish live amid the anemone's tentacles for the safety and for food particles, and they help it by scaring away intruders and perhaps cleaning the anemone up a bit. Other species of anemone hold onto hermit crabs (thus protecting them), with at least one eating the crab's dribbled-down food. One anemone actually makes a golden shell from scratch for its crab to inhabit.

Listen to more of their lovely names now—hell's fire sea anemone, strawberry anemone, snake locks anemone, and, simply, marvelous sea anemone. They all are.

Coyote

SUBURBAN MENACE?

Coyotes are wild canines, a family that includes the likes of wolves, foxes, dingoes, wild dogs, and jackals. Although there are probably more red foxes in all the world than coyotes, these howlers are very common and becoming more so. Coyotes are extending their range every year in North America, and they now live in almost every state and Canadian province; they also roam through Central America. There are nineteen subspecies of coyote, but it is difficult to distinguish among them, since they interbreed quite often now, even with wolves and domestic dogs.

The coyote population is growing because they are commensal with people, sharing our table, so to speak. Where there are farms there are rodents, which attract and sustain the average coyote over the warm season. They also eat berries and other fruit. Coyotes sometimes kill domestic sheep, and sample carrion regularly, along with the occasional live deer or other large ungulate, particularly when these larger prey are weakened by a hard winter. They are able to travel 40 to 50 miles a night, and ready to hunt in early morning and then early evening (usually). For hunting, they rely on sight most often, then smell and hearing, to get their prey. At just 30 to 50 pounds, coyotes usually cannot kill a large, healthy animal, even if they attack together in a pack. Coyotes have also been known to kill human infants, particularly around the edges of suburbia, although this is rare.

The more food there is around, the greater is the tendency for coyotes to live in packs as opposed to alone or in mated pairs. The packs can get quite large and pronouncedly territorial when there is yet more food. These groupings are usually composed of one sexually active pair and their pups from the last few years, all helping to defend the carrion and watching the newest generation of pups grow up. Coyotes mate for life, monogamously, and typically raise six pups at a time in dens in the ground. The puppies look just as cute as baby domestic dogs.

If you hear a yowl or see an unfamiliar neighborhood "dog" that looks a bit scruffy and is a mottled combination of white, gray, brown, and rust, and certainly if you see this "dog" with its head lowered to the ground and mouth open in a threat gape, bring children inside and keep the door shut. Although they are highly efficient predators and are known as tricksters in many Indian legends, they cannot yet open the back door.

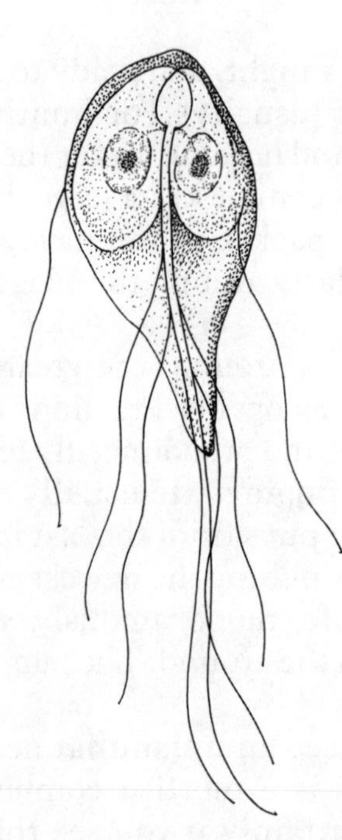

Giardia Lamblia

POWERFUL PROTOZOAN

This is a creature with whom you may be intimately familiar. If you have drunk water from what looks like a clean, pristine mountain stream or used that water for cooking or rinsing without boiling it, if you have sampled the water or food in a country without adequate water treatment, if your child goes to a daycare center where sanitation is not a top priority, or if you were skiing Aspen in the wrong winter in the 1960s . . . then you may well have mingled with *Giardia lamblia*, also known as *Giardia intestinalis*.

Giardia brings with it severe diarrhea, bloating, cramps, and flatulence, all for seven to ten days (which is long for such a drastic and debilitating constellation of symptoms). If untreated, it can quite easily turn into chronic diarrhea. The incubation period, during which it is multiplying inside of the body to reach dramatic levels, is one to two weeks.

The life of this protozoan parasite—only one cell big and shaped like a microscopic pear—is pretty simple. It lives in the surface water in mountain areas and then travels on down the mountainside into the drinking and cooking water in places where water treatment is not adequate (even temporarily). It infects animals, who can pass it to people through their feces (which can, of course, be what is contaminating the water with large supplies of this protozoan). Once a person takes it in, it clings to the walls of the upper small intestine, changing its antigen coat quickly to prevent antibodies from attacking it. There are surging currents in there, rushes of bile detergents, acids, enzymes, and more—but Giardia holds on tightly. From there, it can infect the next person who gets near the first person's feces.

Fortunately there are only three species of Giardia. But while only one of these (our subject here) attacks people, it is the most commonly identified intestinal parasite. It lives all over the world, and even in the United States it infects about 4 percent of the population. In the Third World it is pervasive, one of the first things to infect children (sometimes even through the breast milk of their mothers). People who are malnourished, immunodeficient, or recent recipients of gastric surgery are especially at risk. Many times, they are able to fight it off to a certain extent but become silent carriers of it, infecting others.

Tasmanian Devil

THE NAME SAYS IT ALL

It is probably the yowls, screams, and growls of this animal that frighten people the most. It vaguely resembles a hyena but with a broad, short snout, large teeth, jaws strong enough to crush large bones, and ears that are red on the inside. It is also fast-moving. Nocturnal and solitary, it mostly rests in dens or under brush in the daytime. Its threat gesture is a wide open mouth.

This devil is about 2 feet long, not counting a long tail, and rises nearly a foot high to the shoulder. It is 9 to 20 pounds of fierceness, mostly in black with some large white spots. The last of a group of animals that once specialized in carrion, the Tasmanian devil is a marsupial, a kind of pouched counterpart to the hyena. It carries its babies in the pouch for their first twenty weeks of life, whereupon they climb onto their mother's back and try to stay on.

Once found all over Australia (connected by land bridge to Tasmania until the last Ice Age), the Tasmanian devil then became quite rare. No longer. Since the extinction of the marsupial wolf, its numbers have bounced back well in Tasmania. Look for it there (an island off Australia's south coast) only where it hunts all night through the forests (especially those near the shore). With good hearing and sense of smell, but poor vision, its favorite hunting method is to lie in wait, then surprise the prey. It eats rats, rabbits, chickens, half-grown sheep, even insects and plants (if the killing is going poorly), and also carrion. It gets along well living near people, by stealing their farm animals to eat. It also makes its presence known by yowling fiendishly every so often.

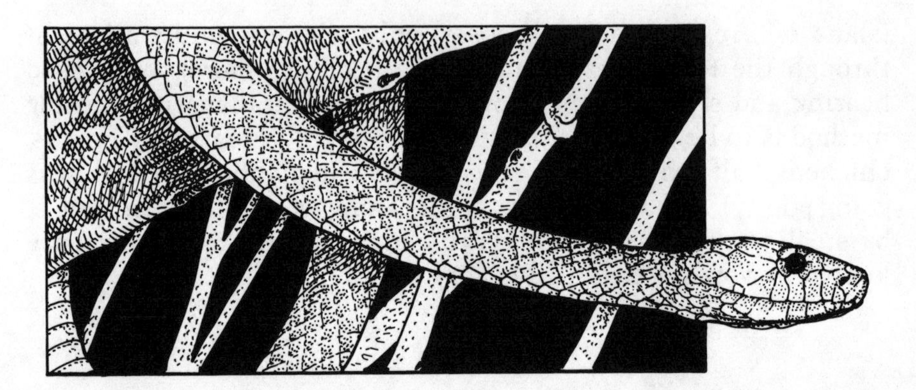

Black Mamba

THE FASTEST SNAKE ON EARTH

The black mamba, along with its relatives the green mamba, western green mamba, and Jameson's mamba, are the speediest snakes on the planet. They have been clocked at about 8 miles per hour, faster than most of us can jog. Long and thin, the aggressive black mamba lurks either in bushes, trees, or on the ground in Africa. It is not really black, but rather outlined in black.

There are stories of black mambas outrunning motorcycles, surviving until sunset no matter when they were wounded, and enjoying a drink of milk. None of this is true, though their speed is certifiable. And, as Aesop wrote, "It is easy to be brave from a safe distance."

Turkey Vulture

A LOCAL CELEBRITY

A strong flyer who soars high, the turkey buzzard can see carrion a mile away and accelerate to 60 miles per hour to get there first, then plunge its head into the dead animal. It uses its quite keen sense of smell only as a backup. The bird's sense of taste is, perhaps mercifully, quite rudimentary.

It may carry some of the carrion back to the nest to feed to its young, in chunks partly digested. It has a stark white bill and tan feet on a large blackish-brown body, and makes for noise only a small grunt. Yet despite its unpleasant behavior, the turkey vulture is a creature dangerous only to the already dead.

While the bacteria from the carrion would be a health problem for most animals, this creature has a digestive system able to destroy the bacteria that surely infest its nasty dinner. Its naked, red, wrinkled head helps too, since there are no feathers where yet more bacteria could become lodged. The bird always cleans its beak after dinner, and also goes to a high, sunny, windy place to stretch out its neck and wings to the sun for more cleaning. It defecates down its own leg after eating, but even its wastes are antiseptic.

Turkey vultures are among the more than 400 birds of prey that roam all the world save Antarctica. Of the two broad groups of vultures, the turkeys are one of the New World vulture species (others include the black vulture, California condor, and the giant Andean condor with a 10-foot wingspan). This New World group, which seems to be related to the storks, evolved its deathly eating habits and related traits separately from the Old World vultures; the latter are related to the eagles. (Carrion is ample reward everywhere, I guess.) The Old World group includes the only tool-using vulture: the Egyptian vulture picks up rocks in its bill and throws them down onto ostrich eggs until they break, getting runny enough for good eating. Another Old World cousin to the turkey vulture is the lammergeier, a huge Eurasian species that can knock down animals like lambs with its wings alone.

The turkey vulture is a cause célèbre in at least one town: Hinckley, Ohio. Here, the citizens await its return every year. On or about March 15, the flock of turkey vultures indeed sweeps into town, each one up to 30 inches long and with a wingspan of up to 5 or 6 feet. They are ready to mate, roost in nearby caves, and lay their two eggs per family.

When you hear about the swallows coming back to Capistrano, think, too, of the turkey vultures returning to Hinckley.

European Viper

TOURISTS, ALERT

A member of the viper family, this slinky creature is close cousin to the puff adder (see page 80), the snout-headed lance head, the bushmaster, the saw-scaled viper, the horned viper, and others. Pit vipers, their larger extended family of 144 species, all live in Europe, Africa, and Asia, and all have two deep pits between the mouth and snout; they use these as heat detectors to help discern both prey and predators. (The other branch of the family includes the rattlesnakes.)

In the United States, about 8,000 people are bitten each year by members of this pit viper family, many of them foolishly teasing the snake they come upon by the side of the road. Most victims do not die, but quite a few end their escapade with black, blistered flesh and permanent paralysis of the body part bitten. The strike lasts just a fraction of a second, but in that time the snake learns the scent of the prey. That way it can find the creature after its death a while later.

European vipers extend unusually far north in Europe, up even to

Norway, where they manage by hibernating for the very long winter. Known for their stubby bodies and arrow-shaped heads, the European viper is the only venomous snake that commonly lives in Britain, where it has, upon occasion, killed small children and adults with preexisting heart problems. Their bodies are darker in the northern part of their range, allowing them to absorb heat when the sun finally waxes warm. In the spring, the males emerge from their winter naps ready to quarrel over the females, who make the scene a few days later. In combat, two European viper males twist their upper bodies up and against each other, each attempting to push the other over. Unlike most snakes, when the female viper gives birth, it is to live young rather than eggs.

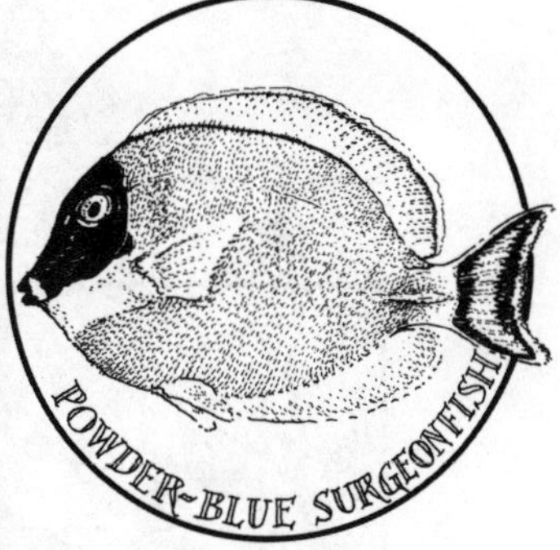

Surgeonfish

PRICKLY SPINES

The surgeonfishes—76 different species strong—graze in large schools near coral reefs in all the warm ocean waters of the world, particularly in the Indo-Pacific. Often colorful, they have small scales, a body outline shaped like an egg, and chisel-like teeth for chomping on algae and other herbivorous fare.

Their name does not come from their teeth, however, but from their tail and fin spines, a sharply formidable defense. As whetted as a surgeon's knife, the tail spine can be shot out to its full length or retracted; its grooves conceal tiny venom glands. The fin spines, which may also hold the small venom glands, are used like slashing bodies when the fish is threatened.

If you get on the wrong side of a surgeonfish, expect lacerations with profuse bleeding and lots of pain, even nausea. Proceed immediately to a hospital. Within an hour, the relevant body part will swell, staying that way for about ten days, and taking a full three weeks to heal completely. The only real treatment is to soak the wound every day in epsom salts and warm water, then to rub on an antibacterial ointment.

Home aquarium owners note: the blue tang is a surgeonfish.

Skunk

BLINDING EFFLUENTS

Because it lacks a sense of smell, the great horned owl is the only real enemy of the skunk. Dogs give chase too, but having an excellent sense of smell themselves, they usually learn their lesson fast. So skunks stroll with an air of considerable confidence. They know they have a trick up their sleeves (so to speak)—from two anal glands, under careful muscular control, each the size of a grape. Out, via the two nipple outlets, comes the musk. With speed and accuracy, the skunk can shoot up to 14 feet, usually aimed for the eyes of whatever is disturbing it. The result can be temporary blindness as well as the familiar stink. The skunky substance is oily, clear amber in color, and made largely of sulphur. Believe it or not, at night it is luminous. It smells terrible both day and night, too.

Look for three warning signs. First, the skunk will stamp its feet.

Second, it will raise its tail, with tail tip pointed down. Third, it will lift up the tail tip, spread out the tail, and bend its body into a "U." This is so that both its face and rear are pointing at you. This aim is followed quickly by its "fire." A skunk can fire five or six times in a row.

Watch for the common skunk anywhere in the United States except for desert areas, also north into southern Canada and south all the way to Guatemala. A nocturnal animal, it also lacks good senses of vision, smell, and hearing, so it may not notice your approach. If you see one, freeze and certainly make no sudden movements.

If you have indeed been skunked, rub three to six large cans of tomato juice onto your skin, or your dog's fur. Wait a few minutes, then rinse. The smell may well come back for a while later, when the victim gets wet.

Strange to say, the skunk's stink was once thought to be a good treatment for asthma. Doctors told their patients to keep some in a tightly sealed bottle, then sniff a bit when their air passages were closing. Perhaps even stranger, people used to eat skunks. And some still have them as pets, once they have been de-skunked.

When a skunk is not "skunking" something, it is shy, easygoing, and usually quiet (though its spectrum of noises includes a churr, a grunt, a squeal, a growl, and a twitter). An omnivore, it ambles through the edges of woods and other areas that are partly open, never more than 2 miles from water. Here it is looking for insects, its favorite food. It will stuff itself on grasshoppers, for example, and dig hard for many insects and grubs. It also likes tiny mammals, such as the mice it can discover in their burrows. Only about a third of its diet is vegetarian, mostly berries, other fruits, and grasses.

A striped skunk's year begins with mating, then denning under a stump, log, abandoned burrow (of another animal), or even your back porch. In April or May, its five to ten tiny kits are born. These babies, which the mother picks up as though she were a cat, don't leave their den until June or July. This is why you see lots of dead skunks on the highways in these months: the newly grown animals are searching for their own territories. When full-grown, the skunk is 24 to 30 inches long and weighs 4 to 10 pounds, with the tail taking up one-third of its length. It eats all its first summer long, dens communally in late fall, then starts the year all over again. In the wild, a skunk lives only one to two years.

Skunks are members of the weasel family, all of which have musk glands. But only the skunk employs them as the weapon of choice.

Goose

CRUISING FOR A BRUISING?

True, this creature hardly ranks in danger with the rhino, sea snake, killer bee, great white shark, or diamondback rattler. But it is tough enough to have been used as a "watchdog" in the Middle Ages. People housed their goose under the stairs by the front door to guard the place—and it would even come when it was called.

The goose can peck hard, if annoyed, thus causing plenty of pain and bruising. Anyone who has ever stood right next to one while throwing bread crumbs knows that they are large birds with strong wings who can stare you right in the eye, hiss meanly from a big beak, and come right after you to wrest away more bread crumbs. Their hissing is a threat gesture, part of their ritualized fighting. Watch out!

Tamagasse

THE JUMPING VIPER

This highly dangerous member of the pit viper family is aptly nicknamed the "jumping viper." When aroused, it hurls itself so strenuously at an enemy that its body actually lifts off the ground on the way. A venomous bite may well follow.

The Tamagasse's venom secretion, like that of other venomous snakes, begins even before it is born. The main venom gland, just above the mouth, produces the substance, which is then stored in a reservoir nearby. As required, this fluid flows down a tube where another gland adds to the brew at the last moment. Then muscles squeeze it out. The snake does not use all of the venom in every strike. In fact, usually only 11 percent to 50 percent squirts out each time, and some strikes feature no venom at all. The snake can replenish an entire supply in between sixteen and fifty-five days, depending on the species and on the temperature, producing more venom on warm days. And there are lots of warm days in South America, where the Tamagasse lives.

Weever Fish

WATCH YOUR STEP

This family of well-armored fish lives safely in deep water—about 150 feet down. Weever fish inhabit the Atlantic from the latitudes of Britain and mainland Europe south to the shores of Morocco. There are many of them in the Mediterranean. When they come up to feed, they bury themselves in the sandy bottom where water depths are just 10 to 20 feet. They are in ambush, looking for food in the form of small fish and crustaceans.

Scuba divers and others should beware of the most salient feature of this otherwise nondescript brown or tan fish: the sharp, separated spines that rise from their backs. These spines are not used in food gathering, but rather in defense. To tread on them is to feel searing pain. Always get medical help, even though the venom emitted from the spines is unlikely to be fatal. More advice from the experts: treat your wound with your own urine first.

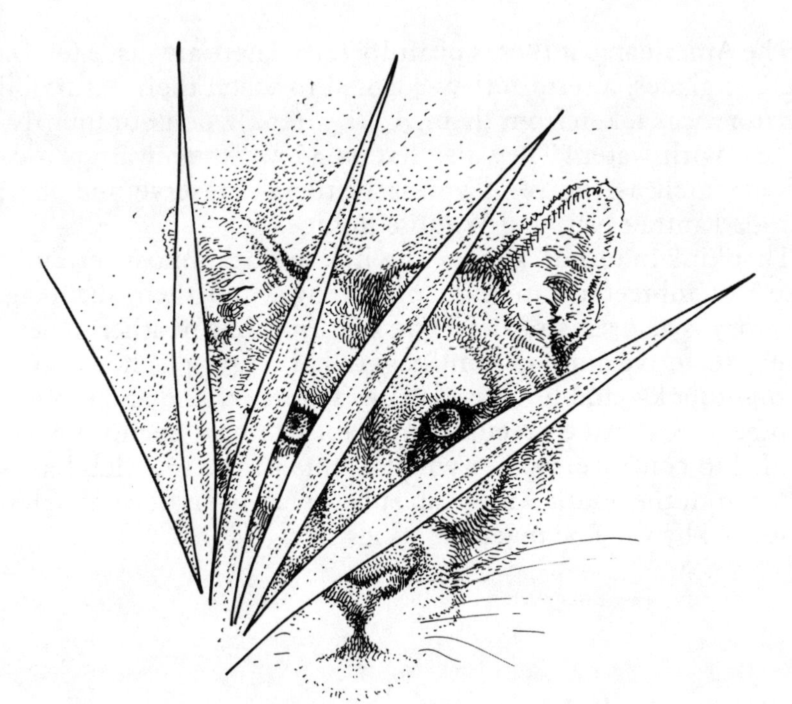

American Panther

BIG CAT

The panthers of India are black, while American panthers are tawny to blackish. The American panther also has a kink at the end of its tail, a cowlick of hair in midback, and white flecking on its head, shoulders, and neck. It is strong and solitary.

Let us get the wildcats of the Americas straight first. Besides the lynx and bobcat, there is a bigger cat, which is known by different names—cougar, puma, panther, mountain lion, catamount, and more—but is essentially the same animal in different territories. The American or Florida panther is, then, one of the subspecies of a powerful animal that once ranged more widely than any mammal in the Americas (from northern British Columbia in Canada, to eastern Nova Scotia, through the United States and Mexico, and down to the tip of South America). Today, all of these big cats are rare to some degree.

The American panther is pitifully rare. There are just a few left in the Everglades, a terrain they adapted to when their natural, drier territory was taken from them. (Cats, after all, don't ordinarily like to deal with water.) They also live in a few other swampy areas of Florida such as the Big Cypress National Preserve and the new Florida Panther National Wildlife Refuge.

That kink in its tail, and the cowlick too, are almost certainly the result of inbreeding in this small range, and their shrinkage of territory is also causing defective sperm among panther males. It is a very secretive animal—unlikely to bother people, however scary it might look—and in desperate need of more land and more sheer numbers. A captive breeding program has now begun but is embroiled in controversy over whether this panther (which has some genes from the South American subspecies) is unique enough to be protected by the Endangered Species Act.

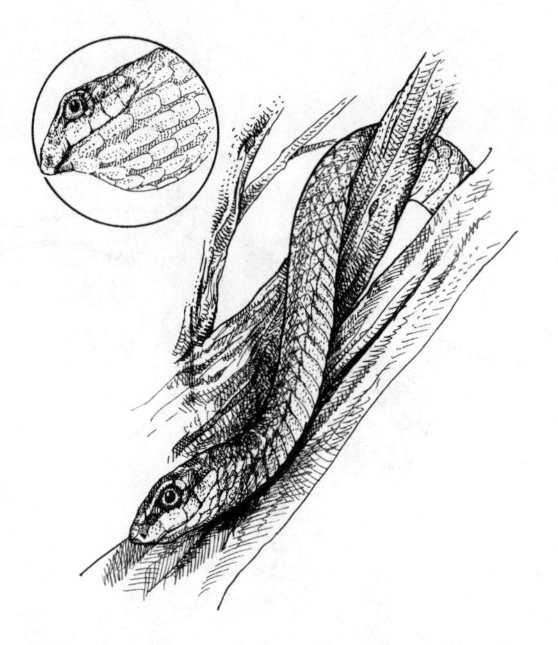

Boomslang

TREE-BORNE TROUBLE

This venomous snake is known for its biting apparatus. The venom is administered slowly down grooves in its fangs that lie at the back of its mouth. The boomslang shares this atypical back-fanged (as opposed to front-fanged) design with three other poisonous snakes that live in Europe—the hooded snake, the European cat snake, and the Montpelier snake—and many in Madagascar, Australia, and the Far East. The latter group includes the flying snakes, which glide by spreading themselves out to catch the air. And a more distant relative bears the charming name of dog-faced water snake.

Grayish brown or electric green, the boomslang usually stays up in its arboreal home. It moves up and down the tree trunk—fast—after its prey of frog or lizard. When it feels cornered, it may hiss at high decibel and puff out its neck to show off the skin among the scales. At this point, anyone nearby should ponder that, though the snake hangs on tight, it usually does not administer a fatal dose of venom.

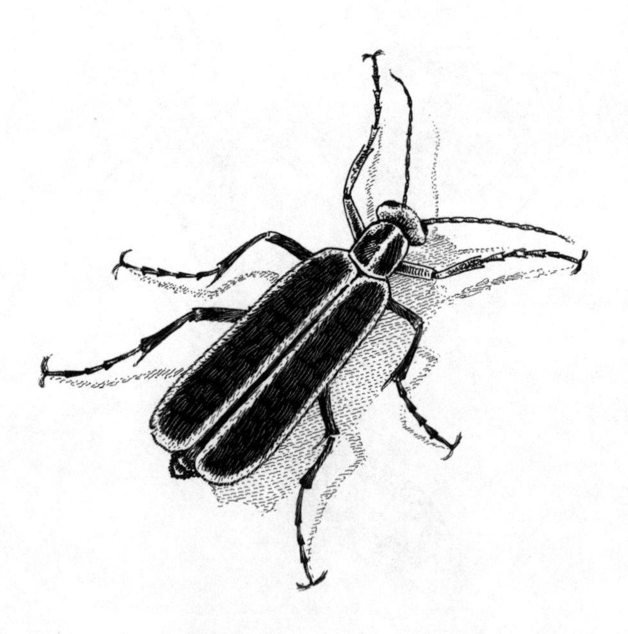

Blister Beetle

COMMON AND PAINFUL

The blister beetle's name says it all—touch one and your finger may well blister, painfully. Even a long-dead, dried, blister beetle has the same effect. The cause of the irritation is the chemical, cantharidin, perhaps the oldest medicine to have been derived from an insect. Hippocrates knew of it in the fourth century B.C., and he recommended the beetle extract as a cure for dropsy. Beginning almost 2,000 years ago, it was prescribed to create blistering, since doctors then believed that damaging substances could be induced to leave the body through blistering as well as bloodletting. It was also used in baldness treatments as recently as the 1940s. And its most famous, though now thoroughly discredited, use is as "Spanish fly." Extracted from only one green beetle from the blister beetle family, this substance was once supposed to be good for everything from gout to rabies to earache, and especially excellent as an aphrodisiac. What it does in the latter dimension is to irritate the genital tract, creating not only pain but harm to the kidneys. Just 30 milligrams of this beetle product can cause death.

The birds may outdo us in wits here, as they simply spit out blister beetles. Even ants wise up fast; when a group of them attacks a blister beetle, and it puts out its defense, they run away to go roll in the grass to get it off.

Even when they are not blistering, these beetles are nasty. In clusters on some plants, where they often are found, they are said to smell a bit like dead mice. They also chow down a lot of crops from alfalfa on down the edible alphabet. They can be quite prolific at all these activities too, since a female usually lays about 1,000 eggs at once. In one of their several larval stages they look a little like lice and attach themselves to the legs of bees and other insects. Their one virtue, from our human point of view, is that some of them eat grasshopper eggs.

But their staying power obviously deserves our respect. Here is how one turn-of-the-century naturalist described the march of insects: "When the moon shall have faded out from the sky, and the sun shall shine at noonday a dull cherry-red . . . and no keels shall cut the waters, nor wheels turn in mills, when all cities shall have long been dead and crumbled into dust, and all life shall be on the very last verge of extinction on this globe; then, on a bit of lichen, growing on the bald rocks . . . shall be seated a tiny insect, preening its antennae in the glow of the worn-out sun, representing the sole survival of animal life on this our earth—a melancholy 'bug'." The survivor might well be a beetle.

Bouga Toad

STRANGE EFFECTS

The venom of the bouga toad, also called the marine toad, has an active ingredient used in potions by the shamans of Haiti to induce a catatonic state. The person who is given the extract may appear to die, then even be buried in a shallow grave, only to arise and do the shaman's bidding while still under the influence of the toad's venom. But this "zombie," or "living dead" person, was actually never quite dead to begin with. Once the venom wears off, the zombie recovers.

Other ingredients in a successful zombie recipe may include the poisonous datura plant and various fillers, but the bouga toad venom has been shown to be the most effective component in the stew. Live toads, which come in most of the colors of leaf litter, are often thrown into the boiling mixture during its preparation so that the venom squeezes easily out of their glands. Marine toad venom was also used—in smaller doses—as a hallucinogen by the Maya and other Central Americans.

Lynx

NORTH AMERICAN WILDCAT

This wildcat will not bother you, unless you happen to be sleeping outside in the woods on its very, very hungriest night. It eats mostly snowshoe hares, also squirrels, gophers, and birds, and it is strong and sneaky enough to kill a deer or a caribou calf. Its sharp teeth go for the jugular. Its killing method: stalk and pounce, or hide and wait. Usually solitary, lynxes will hunt as a team when a mother's kits are nearly fully grown.

Look for it in the forests all over Canada, Alaska, in the northern reaches of Michigan–Wisconsin–Minnesota, and down through the Rocky Mountains to southern Utah. The lynx will go farther south if food is scarce. The animal does not mind high altitudes or very deep snow since its broad paws are excellent snow shoes. Actually very secretive, shy, and mostly nocturnal, a lynx is very hard to see.

This silver-gray wildcat shares North America with the bobcat (who has taken the more southerly woods) whom it closely resembles. The lynx looks less like a house cat than the bobcat does, mostly because of the pointed hair tufts that rise up from the tips of its ears. It also has a thick ruff of hair around its face and a stubby tail. About 20 inches tall to its shoulders, it weighs about 25 pounds. It looks strong, and is.

There are fewer lynx then there used to be, especially in northern Europe where they originated. Some years are better than others, since its population teeter-totters with that of the hare it preys upon so avidly. Those who want to see the lynx thrive should note that they are trapped in Canada and may appear in fur coats.

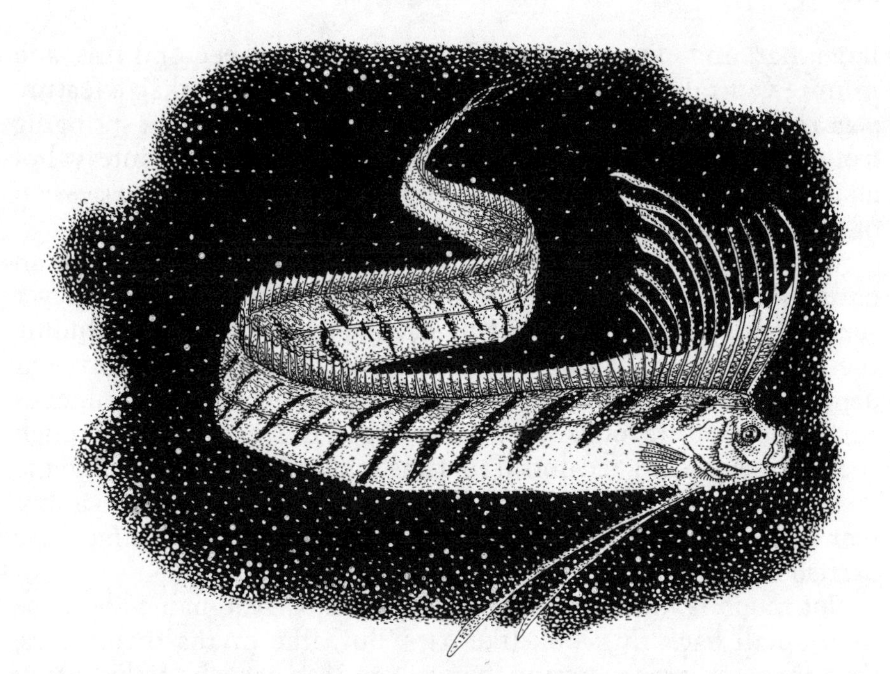

Oarfish

THE ORIGINAL SEA MONSTER?

"Their attention was attracted by a rushing sound in the water," wrote a nineteenth-century sea captain. Sailing off the coast of Bermuda and, "On reaching the spot from which the commotion arose, they discovered a large sea monster that had thrown itself onto the rocks. It appeared to be dying from a frantic effort to reach the ocean water again. The two men grabbed large forks, which were strewn on the beach as tools with which to gather seaweed, and attacked the monster."

"The reptile was 16 feet 7 inches in length," Captain Hawtaigne continued, "tapering from head to tail like a snake, the body being a flattish oval shape, the greatest depth, at about a third of its length from the head, 11 inches. The colour was bright and silvery; the skin destitute of scales but rough and warty; the head in shape not unlike that of a bull-dog, but it is destitute of teeth; the eyes were large, flat, and extremely brilliant, it had small pectoral fins, and minute ventral fins and large gills. But its most remarkable feature was a series of eight long, thin spines of a bright red colour springing from the top of its head and following each other at an interval of about an inch; the longest was in the center: it is now in possession of Colonel Munroe, the acting Governor of the Colony."

It was later confirmed that this "sea monster" was actually an oarfish, an odd creature of the deep ocean. Though almost never seen, it lives in all the world's tropical and temperate seas, including the Mediterranean. Growing up to 35 to 40 feet long, it lives at depths of 300 to 3,000 feet and rarely surfaces. Just a few inches thick, the creature looks like a giant ribbon moving eel-like through the water. A blazing red dorsal fin runs the length of its body, and the crest on its head can stand erect. Unlike a real sea monster, it cannot rear its head out of the water and has no fangs at all. In fact, the oarfish is toothless and feeds by filtering the water.

Not much is known about these elusive oarfish, named because their small back fins look like oars. But, like myths themselves, they are good at regenerating themselves. If an oarfish's tail is bitten off, it simply regrows it.

Their danger to people is slight but emphatic: if one surfaces near you it will not bite or strangle, but it just might scare you to death!

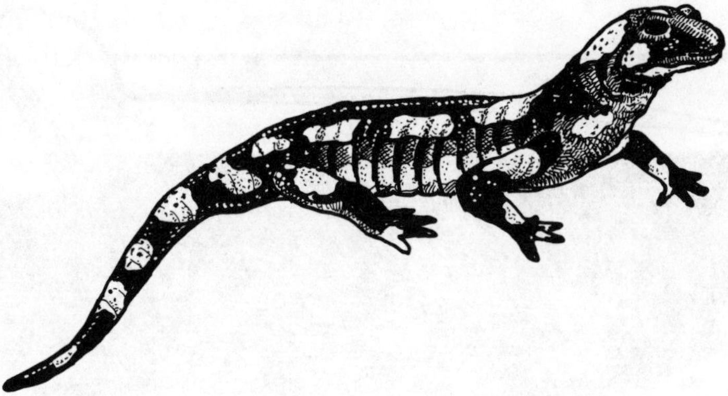

Fire Salamander

FIRE FROM THE FOREST

Like skunks, spitting cobras, some lizards, insects, and other amphibians, the fire salamander can protect itself by emitting a noxious liquid. The effluent can blind you for about twenty minutes if it reaches your eye, where it is usually aimed. The liquid causes plenty of pain in the mucous membranes and will make you very queasy if you swallow it.

Found in the forests of Europe, this black and yellow amphibian is 6 to 8 inches long and can hit you with that neurotoxic liquid from at least 7 feet away. This salamander is easily annoyed—even a blade of grass poked at it will probably make it secrete from the glands in its back. There are many stories about the fire salamander in European folklore. People used to think it could survive fire (hence its name), and that if a woman strapped one to her thigh, she would not get pregnant. Don't bank on the latter.

Porcupine

PRICKLY SELF-DEFENSE

Bearing a name that means "quill pig," each one of these plump, waddling animals can have as many as 30,000 quills. From 1 to 4 inches long, the quills form a stiff nimbus around all of its body, save nose, throat, and stomach. Guard hairs and thicker underfur shelter them when they are not in use. The quills are under muscular control and, contrary to popular conception, need not be flung. When a porcupine is ready to attack, though, it does swing its tail from side to side and chatters its teeth. Then watch out.

An animal with such superb defenses does not run away or easily become nervous. The porcupine sleeps calmly like a dog on the ground or drapes itself over a high tree branch using its long tail for balance. It does have enemies—fishers, bears, wolves, fox, coyotes, cougars, eagles, and great horned owls—but they are often driven away with muzzle, mouth, and eyes full of quills, sometimes to starve to death from the quills in their mouths, while the porkie waddles on. Quills that are "quilled" are replaced in a few weeks of growing.

A word to the wise: even a dead porcupine can get its quills into you and, whether they come from one dead or alive, they must all be removed. If not, the quill tips will literally migrate into your muscles (at the rate of about an inch a day), causing pain and even perhaps temporary paralysis before they migrate out the other side. Very occasionally, a person has been killed by a porcupine when the quills have actually reached vital organs. The best way to remove porcupine quills is with pliers or two objects used as pliers; then use a disinfectant on the wounds. If you see a porcupine, or hear a grunt, whine, or murmur in the woods or woodland clearing, freeze—they don't pay attention to immobile objects. They also don't mind rain or cold or deep of night (with such an overcoat); so look for them in all seasons in their northern forest and occasionally southwestern desert home. (A related species lives in Central and South America.) Sometimes in winter the porkie chews a hole and climbs into someone's summer cabin for the duration. Solitary creatures, they only occasionally den together in the winter, and never truly hibernate.

This is a very successful creature, not stupid at all (in spite of some people's beliefs). At about 18 inches tall and usually 9 to 20 pounds, it is North America's second largest rodent (after the beaver). It uses its hands to eat, a bit like a raccoon. And it can be taught to come when you call it. Some people have even gotten them to shake paws like a dog.

Others have tamed porkies by giving them carrots, lettuce, corn on the cob, apples, oranges, bread, turnips, even potato chips. In the woods, it eats many kinds of plants too, as well as nuts, berries, barks, mistletoe, and pine needles, even flowers like dandelions, clover, lilies, and wild geranium. (In winter in the western states, it kills trees sometimes by eating too much of their bark.) Porcupines also love stuff that is salted, such as leather belts, shoes, and so on, and will even chew on plastic food containers, soap, and ladders.

Last, few people have seen their love dance. In it, they rub noses, touch paws, and walk on their hind feet before mating. Then the porkie has one little baby, weighing about a pound and able to walk the woods within a half hour of its birth. Watch for even those cute baby quills, though.

Gila Monster

VENOMOUS LIZARD

Though there are more than 3,000 species of lizards in the world, and all can bite, only two are venomous: the Mexican beaded lizard and the gila monster.

The poison glands of the gila monster are located at the rear of its jaw, from which the venom runs between its lips and lower teeth, grooved to facilitate the flow. Usually the monster chews the venom into the victim, but it will not bite unless disturbed and in defense. The gila monster is a protected species.

Gila monsters live throughout the southwestern United States and Mexico, and they look a little like fat snakes with four legs. About 2 feet long, they are dressed mostly in black with touches of pink and yellow. Their scales are round and, since they do not overlap, the monster looks a bit beaded. These scales are replaced a few times a year when the lizards shed their skin.

They are always after birds eggs, baby birds, and small mammals, not people, but watch for this "monster" on a desert hike.

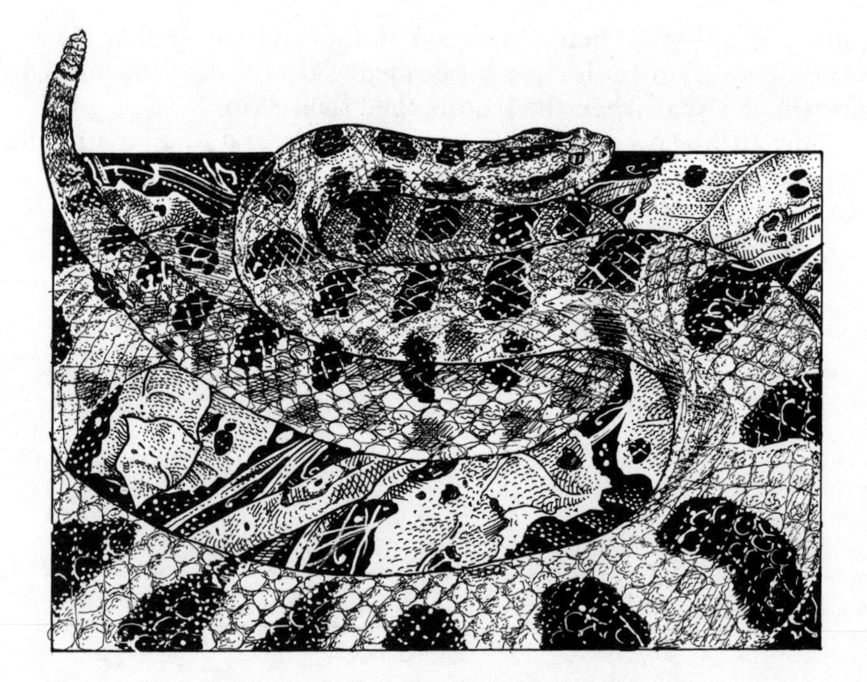

Pigmy Rattlesnake

HARDLY HELPLESS

Baby snakes are generally as venomous as their parents, even when newly born—and they are usually more aggressive too. The pigmy rattler is neither a baby nor even small, and it can be pretty fierce. At 2½ to 4 feet long, this is a snake to avoid. One of the pigmies, called Massasauga, or Great River Mouth, lives in the marshes where rivers end. The other, the true pigmy rattlesnake, or ground rattler, is found on dry land.

While collecting insects, one scientist was bitten on the finger by a pigmy rattler, which he unfortunately confused with a non-venomous snake. When terrible pain segued to bleeding, and then four hours later, to the swelling of his whole arm, he did not become worried. By the time his arm felt very heavy, he called a doctor. He and the doctor agreed that everything was fine, since "the snake was unvenomous." That night, the scientist could not sleep because of the pain and finally got up. He fainted. When he came to, he noticed that his finger had turned black overnight and was dripping blood and pus. This is probably what made him faint the second time.

Forty-eight hours after the pigmy's bite, one whole side of his body was swollen, and the arm on that side was black and blue. But by day three, the swelling and pain were subsiding. By day four, he knew he felt better. By day eight, the swelling had finally disappeared completely. He could not use the bitten finger for another eighteen months, however. Not fatal, usually, but not fun either.

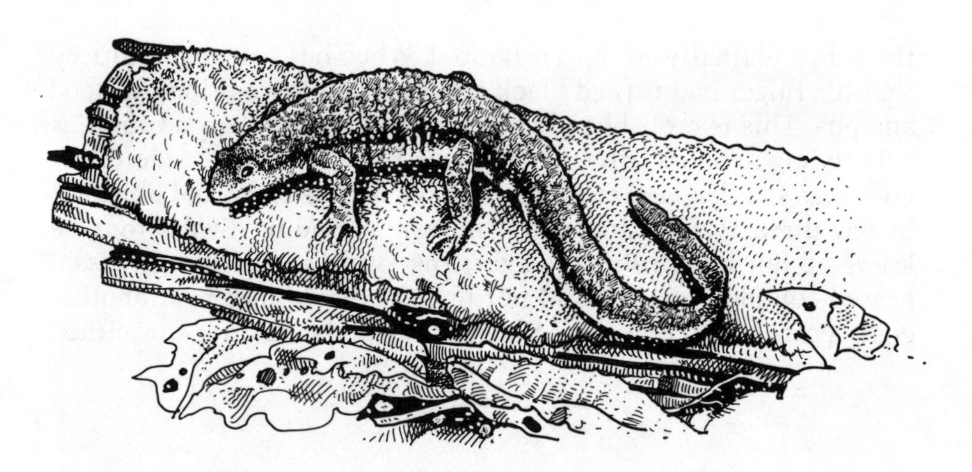

California Newt

RASH VENOM

Although frogs, newts, and salamanders cannot inject their venom directly, in the way that snakes, insects, and spiders do, most of them have some sort of toxic substance to release through their skin in defense. A few of them emit enough of it cutaneously to cause adverse reactions in people, especially when it comes into contact with the eyes, mouth, and sensitive skin. The poison dart frogs (see page 17) of the rainforest are especially toxic, and the cane toad (see page 119) packs a particularly large wallop.

Closer to home are the pickerel frog of North America and the California newt. Neither can kill you, but contact with their skins can give you a rash. The newt means business when it lifts its tail up to show its vivid orange stomach. Don't touch it, and don't let it touch you. Certainly don't hold one over a fire to get more of the toxin to drip out—an old rainforest trick that, with luck, will not catch on in California.

Flying Squid

NOT REALLY AIRBORNE

It is unusual for a squid or octopus to be toxic, but this one is. Fortunately, flying squid live in very deep water, about 1,550 feet down, all day long. Only at night do they swim to the surface to look for food, sometimes breaking the water surface to slide a bit. They bite those in their way. Fisherpeople and late-night swimmers beware.

Bibliography

Attenborough, David. *Life on Earth.* Boston, Toronto: Little, Brown and Company, 1979.

Benyus, Janine. *Northwoods Wildlife, a Watcher's Guide to Habitats.* Minocqua, Wisconsin: NorthWord Press, Inc., 1989.

Berenbaum, May R. *Ninety-nine Gnats, Nits, and Nibblers.* Urbana and Chicago: University of Illinois Press, 1989.

Brooks, Bruce. *On the Wing: The Life of Birds: From Feathers to Flight.* New York: Charles Scribner's Sons, 1989.

Caras, Roger A. *Dangerous to Man.* Philadelphia, New York: Chilton Books, 1964.

Carr, Archie, and the Editors of Life. *The Reptiles.* New York: Time, Inc., 1963.

Coldrey, Jennifer. *The Man-of-War at Sea.* Milwaukee: Gareth Stevens Publishing, 1987.

Costello, David Francis. *World of the Porcupine.* Philadelphia and New York: J. B. Lippincott Company, 1966.

DeJoode, Jon, and Antonie Stalk. *The Backyard Beastiary.* New York: Alfred A. Knopf, 1982.

Department of the Navy, Bureau of Medicine and Surgery. *Poisonous Snakes of the World.* Washington, DC: U.S. Government Printing Office, 1965.

Evans, Howard Ensign. *The Pleasures of Entomology.* Washington, DC: Smithsonian Institution Press, 1985.

Freiberg, Marcos A. *The World of Venomous Animals.* Neptune City, NJ: T. F. H. Publications, Inc., 1984.

Fridriksson, Sturla. *Surtsey.* New York, Toronto: John Wiley & Sons, 1975.

Fuller, Thomas C., and Elizabeth McClintock. *Poisonous Plants of California.* Berkeley, Los Angeles, London: University of California Press, 1986.

Gadd, Laurence. *Deadly Beautiful: The World's Most Poisonous Animals and Plants.* New York: Macmillan Publishing Co., Inc., 1980.

Grzimek, Bernhard, ed. *Grzimek's Animal Life Encyclopedia,* Vols. 1–5. New York: Van Nostrand Reinhold Company, 1968.

Grzimek, Bernhard. *Grzimek's Encyclopedia of Mammals,* Vols. 1–5. New York: McGraw-Hill Publishing Company, 1990.

Halstead, Bruce. *Poisonous and Venomous Marine Animals of the World.* Princeton, NJ: The Darwin Press, Inc.

Halstead, Bruce, et al. *A Color Atlas of Dangerous Marine Animals.* London: Wolfe Medical Publications Ltd., 1990.

Harrison, Kit. *America's Favorite Backyard Wildlife.* New York: Simon & Schuster, 1985.
Harrison, Peter. *Seabirds, an Identification Guide.* Boston: Houghton Mifflin, 1983.
Hendrickson, Robert. *Animal Crackers, A Bestial Lexicon.* New York: The Viking Press, 1983.
Hillard, Darla. *Four Years Among the Snow Leopards of Nepal.* New York: Arbor House/William Morrow, 1989.
Holt, John G., ed.-in-chief. *Bergey's Manual of Systematic Bacteriology,* Vols. 1–4. Baltimore and London: Williams & Wilkins, 1984.
Johnson, Sally Patrick, ed. *Everyman's Ark.* New York: Harper & Brothers, Publishers, 1962.
Klots, Alexander B., and Elsie B. Klots. *1001 Questions Answered About Insects.* New York: Dodd, Mead & Company, 1961.
Lewis, Stephanie. *Cane Toads: An Unnatural History.* New York: Dolphin/Doubleday, 1989.
Lineaweaver III, Thomas H., and Richard H. Backus. *The Natural History of Sharks.* Philadelphia and New York: J. B. Lippincott Company, 1970.
MacDonald, Dr. David. *The Encyclopedia of Mammals.* New York: Facts on File, 1984.
McCarthy, Colin. *Reptile.* New York: Knopf, 1991.
McFarland, David, ed. *The Oxford Companion to Animal Behavior.* Oxford, New York: Oxford University Press, 1987.
McNamee, Thomas. *The Grizzly Bear.* New York: Alfred A. Knopf, 1984.
Mandell, Gerald, et al., eds. *Principles and Practice of Infectious Diseases.* New York: Edinburgh, London, Melbourne: Churchill Livingstone, 1990.
Mattison, Christopher. *Frogs and Toads of the World.* New York: Facts on File Publications, 1987.
Milne, Lorus J., and Margery Milne. *Insect Worlds.* New York: Charles Scribner's Sons, 1980.
Nardi, James B. *Close Encounters with Insects and Spiders.* Ames: Iowa State University Press, 1988.
Nelson, Joseph S. *Fishes of the World.* New York: John Wiley & Sons, 1984.
Page, Jake, and Eugene S. Morton. *Lords of the Air, the Smithsonian Book of Birds.* Washington, DC: Smithsonian Books, 1989.
Peterson, Roger Tory, and the editors of Time-Life Books. *The Birds.* New York: Time-Life Books, 1971.
Rabinowitz, Alan. *Jaguar.* New York: Arbor House, 1986.
Ramsay, Malcolm. "Polar Bears in Hudson Bay and Foxe Basin," in *Canadian Inland Seas.* Amsterdam: Elsevier Science Publishers, 1986.
Reader's Digest's *ABC's of Nature.* Pleasantville, NY: The Reader's Digest Association, Inc., 1984.

Ritchie, Carson I. A. *Insects, the Creeping Conquerors.* New York: Elsevier/Nelson Books, 1979.

Ross, Charles A., consulting ed. *Crocodiles and Alligators.* New York and Oxford: Facts on File, 1989.

Scott, Jack Denton. *The Duluth Mongoose.* New York: William Morrow, 1965.

Shapley, Deborah. *The Seventh Continent, Antarctica in a Resource Age.* Washington, DC: Resources for the Future, 1985.

Stafford-Deitsch, Jeremy. *Shark: A Photographer's Story.* San Francisco: Sierra Club Books, 1987.

Stevens, John D., ed. *Sharks.* New York: Facts on File Publications, 1987.

Stewart, Darryl. *The North American Animal Almanac.* New York: Steart, Tabori & Chang, 1984.

Stirling, Ian, Charles Jonkel, et al. *The Ecology of the Polar Bear (Ursus maritimus) Along the Western Coast of Hudson Bay.* Ottawa: Canadian Wildlife Service, 1977.

Sweeney, James B. *A Pictorial History of Sea Monsters and Other Dangerous Marine Life.* New York: Bonanza Books, 1972.

Thapar, Valmik. *Tiger, Portrait of a Predator.* New York: Facts on File, 1986.

Van Wormer, Joe. *The World of the Moose.* Philadelphia and New York: J. B. Lippincott, 1972.

Vandenberd, John. *Nature of Australia.* New York and Oxford: Facts on File, 1988.

Westbrooks, Randy G., and James W. Preacher. *Poisonous Plants of Eastern North America.* Columbia, SC: University of South Carolina Press, 1986.

Wheeler, Alwyne. *The World Encyclopedia of Fishes.* London: MacDonald & Company (Publishers), Ltd., 1985.

Young, Steven B. *To the Arctic.* New York: John Wiley & Sons, Inc., 1989.